AF886612

A Boy Naturalist in the Amazon:

The Travels of Ernest T. Morris

Jerry Kuntz

Wickham House

Warwick, New York

Copyright © 2017 Jerry Kuntz

All rights reserved.

Wickham House

30 Wickham Drive

Warwick, New York, 10990

ISBN: 0692927646
ISBN-13: 978-0692927649

CONTENTS

ACKNOWLEDGMENTS

The author's initial ignorance on nearly every topic involved in telling the story of Ernest T. Morris requires a lengthy list of thanks for those who helped to provide context and otherwise assisted in the production of this text. In no special order, I'm indebted to the following:

Eduardo Brondizio; Jeremy M. Campbell; Brian Ferguson; Janet Chernela; Bruna Rocha; Helmut Schindler; William Balée; Laila Williamson; B. Ricardo Brown

Brittany Ray; Joe Jackson; Eisha Neely; Donald Worster

Wayt Thomas; Esther Jackson; Allison Munsell Napierski; W. Douglas McCombs; Paul Grondahl; Phoebe Bender; Karen C. Beck; Susan Leath; Sandra Waxman; Timothy McCabe; Cathryn Anders; Maureen O'Brien; Patrick Raftery; Tim Garrity

Charlotte R. Morrill of The Southwest Harbor Public Library Digital Archive

Margaret Bridge of Black Meadow Flora

Will Higgins; Mike Perkins; Eric L. Mundell, all of Indianapolis

Martin Lubikowski of ML Design

...and most of all to Marianne Lanman Salaymeh, for allowing me the privilege to get to know her great-great-grandfather.

EPIGRAPH

Boy reader, I am told that you are not tired of my company. Is this true?

"Quite true, dear Captain,--quite true!" That is your reply. You speak sincerely? I believe you do.

In return, believe me, when I tell you that I am not tired of yours; and the best proof I can give is, that I have come once more to seek you. I have come to solicit the pleasure of your company--not to an evening party, nor to a ball, nor to the Grand Opera, nor to the Crystal Palace, nor yet to the Zoological Gardens of Regent's Park,--no, but to the great zoological garden of Nature. I have come to ask you to accompany me on another campaign--another grand journey through the fields of Science and Adventure. Will you go?

"Most willingly--with you, dear Captain, anywhere."

Come with me then.

--Captain Mayne Reid, *The Forest Exiles; or, The Perils of a Peruvian Family Amid the Wilds of the Amazon, (1855)*

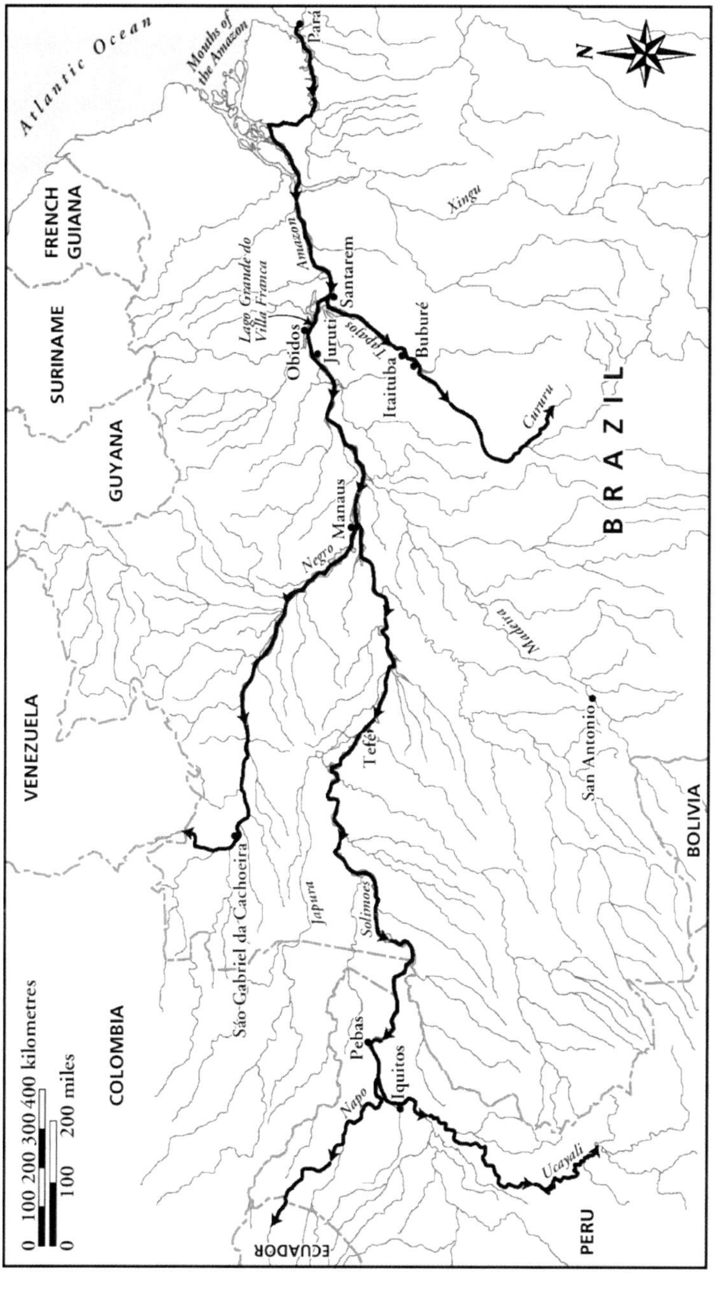

The Amazon Travels of Ernest T. Morris

1 THE STANLEY OF SOUTH AMERICA

In the month of March 1916, Dr. C. Hart Merriam resolved to dispose of his head. Not the one on his shoulders, but rather a head that had been in his collection of biological specimens since he had been a medical student at Columbia University, forty-one years earlier. Merriam had gone on to earn his medical degree, but quickly abandoned a career as a surgeon to pursue his first passion, the study of natural history, including ethnology, the study of different peoples of the world. As a biologist, he advanced through different positions working for the United States federal government, culminating with his appointment in 1900 as head of the U.S. Biological Survey, the predecessor of the U.S. Fish and Wildlife Service. In that role, he was charged with enforcing the Lacey Act, the first federal legislation outlawing trade in illegally harvested wildlife. Merriam enjoyed debating the subject of animal classification with his Washington, D.C. neighbor, President Theodore Roosevelt.

The head in Merriam's personal collection was a dried human head, a trophy made by a tribe of warriors--but it was not associated with any of the North American peoples that Merriam had studied. He had tried to influence Teddy Roosevelt's thinking on the treatment of American Indians, as Dr. Merriam believed their cultural heritage should be preserved. This particular artifact had come from Brazil, where native populations and traditions had been diminished, suppressed, and assimilated only somewhat less ruthlessly than in North America.

Dr. Merriam, who had learned to respect indigenous cultures and their crafts, came to determine that this object belonged in a museum rather than remaining longer as a curiosity in his private collection; and so, he decided to gift it to the Smithsonian Institution's National Museum of Natural History. Though perhaps reluctant to accept an object composed of human remains, the National Museum could hardly refuse a bestowal from a man praised as "the last of the naturalists," who was also one of the founders of the National Geographic Society.[1]

Merriam's donation was announced in newspapers around the country. Many of these accounts also printed his tale of how he had acquired the relic. Four full decades of memories from a vigorous life, along with a penchant to be an engaging raconteur, caused Merriam to misrepresent some details of that story--and fabricate others. The *Trenton Times* newspaper, among others, recited Merriam's melodramatic version:

> Doomed by His Tattooed Face: Young American Explorer's Hard Luck as Result of Indian Tribal Custom
>
> Captured more than four decades ago by South American Indians and kept prisoner by them for several years, Ernest T. Morris, a young explorer, contrived eventually to escape. He made his way to New York, but found life there unpleasant, owing to his conspicuousness, the Indians having tattooed his face after the tribal custom. The pigments would not come out, and the unhappy young man was unable to raise a beard. So the unfortunate young explorer went back to South America, and, nothing having been heard from him since, he and his pictorial face presumably are still there awaiting death, the great eradicator. This might readily be taken for highly fanciful fiction were it not that the Smithsonian Institution tells it in explaining how the National Museum came by a smoke-dried head of a Brazilian Indian now on exhibition there.
>
> This gruesome object is covered with tightly stretched yellow skin and its eyes are covered with wax cones ornamented with two small rodents' teeth to imitate the pupils. The Indian who originally wore the head was killed by some member of the Parentintin [Parintintin] tribe who used to carry it as a trophy, a string for that purpose emerging from the mouth. Black hair hangs loose from the crown and the ears are decorated with cord insertions and tufts of red and yellow parrot feathers, worn by the Indian in life. The head was presented to the museum by Dr. C. Hart Merriam, of Washington, who tells the story of its acquisition by him.
>
> About forty years ago, Dr. Merriam was approached, he says, by the young explorer, who had a number of the dried heads, and,

> being out of funds, wanted to dispose of them. The owner of the heads narrated a tale of capture while traveling near the headwaters of the Tapajós River, and the attempt by the Indians to make of their captive an Indian like themselves by tattooing his cheeks and lower jaw. For a long time, the explorer said, he was in fear of his life, for the Indians threatened to kill him for his head, to be hung from the lodge poles as a trophy. The explorer was spared, for some unknown reason, and was unable to escape until after two or three years, when the watch became a little lax. By traveling only in the night and hiding by day, the fugitive succeeded in reaching the city of Pará, where he secured passage to New York.
>
> "Arriving in the great metropolis," says the bulletin, in telling the story, "he found himself stranded financially, with nothing but his ghastly heads as assets, and at the same time embarrassed by the very noticeable tattooing with which his Indian captors had decorated his face. Unfortunately, he could not raise a beard, and he would not appear in public as he was. Finding that Dr. Merriam was in the city, and knowing him to be a scientist, he decided to approach him in an endeavor to raise some money on the heads. Dr. Merriam says that they found no difficulty in doing so; the American Museum of Natural History taking two and himself one. He does not remember where the others went; but recalls that all of them were sold at $75 each, netting the explorer about $600. With the receipt of the money, Morris's spirits rose again, but having no place where he could seek seclusion from the public eye, he immediately embarked on another trip to South America, and so far as Dr. Merriam knows, has never been heard from since."[2]

Much of the account that Merriam provided was pure fiction, but this much was true: there once had been a young explorer who brought back dried heads from the Amazon, sold them for money, and used the proceeds to immediately return to the rainforest. In fact, he returned five more times, completing a total of seven trips to the Amazon, his last journey taking place in 1884. Had Merriam not mentioned the young adventurer's name--Ernest T. Morris--it might have been difficult to tell whether any parts of the story were true; because Morris, once publicly hailed as the intrepid "Boy Naturalist" and special correspondent of the *New York World*, had long since faded from public memory.

One of the few who read the announcement of Merriam's donation and recognized Morris's name was Jeremiah Johnson, a former federal judge in Oklahoma, who by 1916 was living in Washington, D.C. Johnson, while still a law student in Indianapolis, Indiana, in the late 1870s, had known Ernest Morris. In fact, he had traveled with him as far as the Brazilian port of Pará (now known as Belém) on Morris's fourth trip to the Amazon in 1878. Johnson knew that Ernest Morris had ventured to the great river basin at least twice after he had returned to New York with the dried heads, and suspected that other details of Dr. Merriam's account were falsehoods. Johnson wondered what had become of Morris--he had last heard of him not long after they parted company in Pará. Out of curiosity and nostalgia, Jeremiah Johnson sent off a query to one of the major newspapers of Morris's hometown, the *Indianapolis Star*.[3]

Upon receiving the query from Jeremiah Johnson, someone in the *Star* office recognized the name of Ernest T. Morris--and knew exactly where the truth about his fate could be tracked down. A reporter was dispatched to interview Mrs. Kate Perry Morris, the widow of Ernest T. Morris. She lived in the same house on North Delaware Street that she had once shared with the young explorer; but she also spent time keeping house at the apartments of her elderly father, John C. Perry, head of J. C. Perry & Co., wholesale grocers, a notable Indianapolis firm. By the time the *Star* interview took place, Kate Morris had seen a copy of C. Hart Merriam's account, and welcomed the opportunity to set the record straight. The first confusion that Kate addressed was the fate of Ernest Morris:

> ...all these wild tales about his being lost in the jungle are untrue. He died at our home on North Delaware Street April 28, 1891, after trying to live away from South America. He couldn't do it. That life meant everything to him, and if he had continued in his work he would have been the Stanley of South America...he tried to settle down to mercantile life, but it did not agree with him.

Kate's reference to Henry Morton Stanley invoked the legacy of the most intrepid newspaper-sponsored explorer of the nineteenth century. Stanley, writing for the *New York Herald*, achieved fame in the early 1870s after locating British physician/explorer Dr. David Livingston in Central Africa. Livingston had become ill and had lost contact with his British associates. Stanley later traced the course of the Congo River from its headwaters to the sea.

Kate Perry also scoffed at the story of Ernest having been captured and tattooed, and the implication that he had escaped captivity with stolen head trophies:

> Recent accounts say that he was held prisoner and tattooed by the natives. This is not true. He was never tattooed in his life. The Eastern papers also tell how he returned to South America because he could not raise a beard to cover the tattoo marks. But I can attest that when he came out of the jungle he had hair down to his shoulders and a beard about a foot long!

Kate denied that he had been held against his will, or that he spent years with any tribe, or that he had absconded with the heads: "He bought the eight heads from the natives, giving them fish hooks, beads, and razors. I first met Mr. Morris at the old Meridian Street Church, where he was lecturing with his heads," she continued, and pointed out that Morris frequently gave talks all around Indiana, a fact that also served to dispel any notion that he was ashamed to show his face. She elaborated:

> He had eight of the heads. His favorite trick was to set them up in his room in the hotel when he reached a town and then leave them. He always threw a cover over the gruesome objects, knowing, of course, that curious maids would investigate what lay under sooner or later. Mr. Morris got more laughs out of their screams than any other sport.

With little hesitation, Kate Perry Morris also related that a gulf that existed between her and her husband. It became evident on their honeymoon. "He took me down to Pará in 1882, immediately after our marriage; but after one trip, I never wanted another. The jungle was impassible and the Amazon River was so rough that several times when we were out in our canoe I thought we would never get back to our hut alive." Despite this admission, Kate Perry Morris did not express that she felt any responsibility for keeping her husband from his passion. Perhaps there was more to the story about his last years that was left unsaid.

After dismissing all of Dr. Merriam's melodramatic elements of the history of her husband Ernest, Kate blithely added her own stunning assertion. She told the reporter, "Ernest wrote a book about his travels, but it was never published. Before he could complete the arrangements for the publication he died. I tried three times to finalize dealings with publishers, but was stricken down each time. I believe that there is an Indian superstition that it is never to be published, and I am convinced that it is true."[4]

Kate Perry Morris's suggestion that Ernest's manuscript carried a South American Indian curse was one way for her to explain his descent into obscurity. His unpublished book draft, along with his clipping scrapbooks and travel diaries, were subsequently handed down and preserved by generations of Morris's descendants, and seen by no one else. Those documents, along with other contemporary records, have been collected together nearly 140 years later and reveal the life of an extraordinary individual. His career illustrates a unique period in scientific history: an era when the fieldwork of collection and observation was influenced by Romantic ideals; when self-taught amateur naturalists rubbed shoulders with the academically trained; when Western nations vied to discover and exploit the natural resources of less-developed nations; and when aboriginal populations and their native lore were facing extermination, enslavement, and assimilation.

By standard measures of scientific accomplishment, Ernest Morris's career had little import. He did not map any uncharted areas; he wrote no scientific papers, nor did he put forth any new theories. He was not an illustrator, or a photographer. He may have collected a handful of new species, but those claims were not well-documented. The specimens he collected were not handed over to universities or experts in classification, but instead were sold to wealthy, acquisitive patrons for their showcase displays. He ventured into native villages and observed their rituals and crafts, but was not a trained ethnologist. Where Morris excelled was in his ability to convey the experience of living in and wandering through a vast, unknown, often dangerous, but overpoweringly beautiful wilderness. No contemporary American readers saw his unpublished book; but they did follow his numerous travel columns in the pages of the *New York World* newspaper. The impact of those accounts, as far as Morris's fame was concerned, was fleeting. The *New York World* was sold to Joseph Pulitzer in 1883, who then swept out editors and columnists of the former regime, including Morris.

Lacking other sources of financial and academic support, Morris's career was channeled into a fascinating byway of natural history—the collection and cultivation of orchids. This guidance was provided to him by two of the most interesting characters in American horticulture: a fugitive, disbarred Boston Brahmin named Edward Sprague Rand Jr.; and Erastus Corning Jr., heir to one of the largest fortunes in America. Their names, along with that of Ernest T. Morris, are virtually unknown today. It is hoped that revealing these personalities--individually and in their interactions with one another--to an unprepared readership will be somewhat like leaving a cloaked display of decapitated heads for the maids to discover.

---

[1] Merriam's career is summarized in: Sterling, Keir B. 1977. *Last of the naturalists: the career of C. Hart Merriam*. New York: Arno Press.

[2] "Doomed by His Tattooed Face," *Trenton Times*, April 30, 1916. Many of the same falsehoods can be seen in "Perils of an Explorer," *Logansport Pharos*, May 30, 1916.

[3] "Recalls Display of Indian Heads," *Indianapolis Star*, April 9, 1916, p. 7.

[4] Ibid.

2 THE GALLANT *LUCY*

Ernest Turner Morris, born in 1856, had a notable paternal heritage, including several generations of American doctors, lawyers, clergymen, and patriots. His great-great grandfather, Isaac Coult Morris, was a Presbyterian minister in colonial Pennsylvania. Ernest's grandfather, Bethuel F. Morris, was a veteran of the War of 1812; and also a lawyer; newspaper editor; surveyor; real estate agent; and bank cashier. Bethuel F. Morris and other members of the Morris family moved from Pennsylvania to Indiana in the 1820s. Bethuel was also widely known as a judge on Indiana's Boone Circuit Court. In that role, he was responsible for ruling that Indiana would not recognize slaves as property when they traveled through the state with slave owners; but would instead consider them freemen.[1]

While many of Bethuel F. Morris's siblings also migrated to Indiana and became successful, they did so more on their own initiative than through inherited wealth. Bethuel had two sons, Samuel V. Morris and Sanford L. Morris. Samuel V. Morris, a lawyer, married into the prominent Harrison family--not once, but twice--after his first wife died he married her sister. These two sisters were the grand-daughters of one United States president, William Henry Harrison; and the sisters of another president, Benjamin Harrison.

It was Bethuel F. Morris's other son, Sanford L. Morris, who fathered Ernest T. Morris. Sanford, unlike his brother Samuel, was not particularly successful in business, but instead found his passion as a Presbyterian lay preacher. Sanford Morris married three times, having had the misfortune to bury his first two wives. His first marriage was to Elizabeth Ferguson, whom he had wed in Indianapolis in 1848. She died just two years later, and had no children. Her family background is unknown, although there were many Ferguson families near Indianapolis in that era. Sanford Morris remarried a year later, in 1851, to another Elizabeth--Elizabeth Turner. Again, nothing is known about her background other than her common surname. In the mid-1850s, Sanford Morris and Elizabeth Turner left Indiana and attempted to settle on land near Georgetown, Texas. It was there that Ernest Turner Morris was born on July 22, 1856.[2]

The Morris family's Texas homesteading effort was short-lived. By 1859 they had returned north, stopping first for a year in Cincinnati, Ohio; and, by 1861, arriving back in Indianapolis. Along the journey north to Indiana, Ernest's younger brother Frank C. Morris was born. Sanford Morris then took a series of sales clerk jobs, earning just enough to clothe and feed his family, and allowing him spare time to satisfy his calling to be a lay preacher. After finishing his workday, Sanford Morris would go to Indianapolis's South Side, where he conducted prayer meetings and song services among the poor--even if few heeded his well-intended efforts.[3]

Ernest's mother and father must have been aware that their son developed an interest in natural history at a young age. In 1868, when he was only twelve, Ernest made a sole venture west from Indiana to Illinois to collect new insect specimens for his collection. When in his early teens, he learned the art of taxidermy, which, along with illustration, was the primary research tool of naturalists of the era. When Ernest was seventeen, glass cases of his stuffed bird collection were put on display in the Fourth District public schools of Indianapolis.[4] As a public-school student himself, Ernest was well-read, but chafed at deskwork, and yearned to be outdoors in the wild.

Elizabeth Turner Morris, Ernest's mother, died of an illness in April 1871, when Ernest was fourteen and his brother Frank was twelve.[5] Her early death was likely the cause of a legal action between the boys and their father over a piece of property. The exact sequence of events is not known, but most likely Elizabeth Turner Morris was named as the beneficiary in some other Turner family member's will; and since she died before that will was amended, her Turner family then decided that her share should all go to her boys, rather than to her husband Sanford Morris. The civil court decided to partition the contested property, giving some of the property to Sanford Morris and other parcels to his sons.[6] A guardian was appointed to oversee the settlement for the underage boys, which likely did not amount to any great sum.[7] It is worthwhile to note that all throughout his adult life, from the age of eighteen upwards, Ernest had a close relationship with his father. He visited him often when he was living in Indiana; and it was to Sanford that Ernest addressed most of his letters from South America.

Ernest's fascination with natural history coincided with the enormous interest in that subject that had been occurring in both America and Europe, reaching a peak between 1820 and 1880. "Natural history" was a term that included several scientific disciplines: biology, botany, geology, zoology, anthropology, paleontology, etc. These sciences were not yet professionalized, and therefore amateurs, i.e. "backyard naturalists," could sometimes make new observations and discoveries. As a hobby, the study of nature was considered a moral recreation, in that it fit in with the concept of natural theology--the notion that the natural world reflected the glory of God's works. Moreover, it was a pastime that was available to all classes, requiring little expense. In Ernest Morris's case, the low budget needed for nature study might have been a major factor in his interest.

His passion was fueled by writings of the famous American naturalists of the early nineteenth century: John James Audubon's *Birds of America;* Thomas Say's *American Entomology* and *American Conchology;* John Torrey and Asa Gray's *Flora of North America*; and, most of all, Louis Agassiz's *A Journey in Brazil.* Despite his extensive reading--and perhaps because he realized he lacked the financial wherewithal to continue to college--Morris ended his academic life at age seventeen, without completing high school. He found employment in early 1874 as a clerk in the offices of the Indianapolis & St. Louis Railway.

It did not take long for Ernest Morris to tire of life as an office clerk. He began to form a plan to embark on a naturalists' expedition of his own: to gather specimens, take notes, and afterwards sell the items collected. As his target, he selected the sub-tropical United States--an area of the country whose flora and fauna were not as well-cataloged as the northeast, and containing specimens sought by northern collectors. Always conscious of his limited funds, Morris decided that he could travel on his own boat all the way from Indianapolis to the Gulf of Mexico. He constructed his craft: a skiff--essentially a flat-bottomed canoe--twenty feet long and just short of five feet wide at its center. He designed an awning for the back third stern section to provide protection from sun and rain; and included a small sail that could be set if the wind was favorable. Morris christened his ship *Lucy,* and announced to friends and family that intended to sail down the White, Wabash, and Ohio Rivers to the Mississippi, and follow it to New Orleans. From that port, he intended to either take a steamer to Jacksonville, Florida; or take *Lucy* along the Gulf Coast to Cedar Key, and then travel via rail to Jacksonville. At Jacksonville, he would go up

Florida's longest river, the St. Johns.[8]

The intended destination of Cedar Key suggests a coincidence too great to ignore. Six years earlier, a thirty-year old sawmill worker living in Indianapolis decided to make a thousand-mile walk from Indianapolis to Florida, and from there on to Cuba and South America. His goal once he reached Florida was Cedar Key, a small village of just a few hundred people, but also the terminus of Florida's only rail line, leading to Jacksonville. His name was John Muir. Ernest Morris would have only been eleven when Muir made his trek, but it is possible that Ernest's father Sanford Morris might have met the man who later became the nation's greatest naturalist, and that led to inspire Ernest to make his own journey.

Ernest must have known he was not fit enough to walk that far, and chose to travel by water instead. Though the Ohio Valley's waterways were crowded with all types of vessels, from majestic steamboats to rafts, the idea of such a long voyage by an eighteen-year-old for the sake of collecting bugs and birds was a novelty. Another inspiration for Morris may have been the well-publicized 1872 canoe expedition by Julius Chambers from the headwaters of the Mississippi down to Quincy, Illinois--about half the length of the river. Chambers reported on the tour with his canoe, the *Dolly Varden,* to the *New York Tribune*. Chambers first ventured to Lake Itaska, the then-recognized source of the Mississippi, but found that it was fed by an even smaller body, Elk Lake, which Chambers named Lake Dolly Varden. Chambers was cited as the leading American proponent of canoeing for recreation, a craze that had started in England with John MacGregor and his canoe *Rob Roy*. An Indianapolis newspaper article announcing Morris's trip compared him to MacGregor; and, as will be seen later, through a friend mentioned below, Morris also had a tenuous connection to Julius Chambers' expedition at the headwaters of the Mississippi.

Most people in Indianapolis who heard about the planned trip did not believe Morris would get very far. Ernest T. Morris, at eighteen, still had the physique of a boy. He was 5'6 1/2" tall and weighed between 80 and 125 pounds (according to different accounts). Moreover, he suffered from bronchitis, and was the office-clerk son of a store-clerk father. However, his enthusiasm and confidence infected those whom he told about the trip, sufficiently to convince a friend of the same age, Barrett Channing Paine, to go along with him. Paine's father, Daniel L. Paine, was a poet and editor of the *Indianapolis Evening News.*

On August 25, 1874, Morris, Paine, and a third man--a sales clerk named Porter from the Trade Palace clothing store--climbed into the *Lucy* and shoved off into the White River just below the wooden Washington Street Bridge in Indianapolis.[9] Porter only intended to accompany Morris and Paine as far as the Ohio River. The skiff was laden with the three young men and their provisions: bedrolls, guns, fishing gear, clothing, and food for a couple of weeks. They paddled down the West Fork of the White River, frequently having to portage *Lucy* past shallows of low water, typical of the West Fork in late summer. Thirty miles south of Indianapolis, near where Indian Creek enters the West Fork just below Martinsville, Indiana, the vessel hit a snag and overturned. The young men were dumped into the water, along with all their equipment.

They lost much of their baggage, and Ernest lost one of his rifles. The disaster was too much for Paine and Porter; they gave up on the effort and returned to Indianapolis. Ernest patched the minor damage to *Lucy* and continued down the West Fork alone. It was difficult to drag the skiff over the rocks by himself, but Morris persevered. The *Indianapolis Journal* later commented, “His ambition was stronger than misfortune.” The situation improved once the West Fork was joined by the East Fork of the White River near Rogers, Indiana, about 130 miles from Indianapolis. The paddle down the remaining stretch of the White River, until it reached the Wabash River, was uneventful. Morris stopped for a week in the nearby town of Mt. Carmel, Indiana. He later said the stop was made to collect bird specimens, but he also needed to replace some of his lost cooking gear.[10]

It is hard to discern what Sanford Morris thought about his son’s exploits. One week after Ernest Morris left Indianapolis, Sanford Morris married for the third time, to a woman named Lydia Bond. She was 24 years younger than Sanford, and only six years older than Ernest. Ernest had been planning his Mississippi journey for months, so it seems unlikely that he left town to avoid seeing his father remarry. Instead, it appears that Sanford Morris took advantage of his son’s prolonged departure to avoid any objections that Ernest might have to the nuptials.

After five weeks, Ernest reached Cairo, Illinois, on the Mississippi, having successfully navigated both the Wabash and Ohio Rivers. From there he sent a letter to his father (likely intended to be shared with the *Indianapolis Sentinel*) describing his journey thus far. He wrote in the style of travel accounts of that era, with an emphasis on sights, sounds, and general human activity; but with virtually no remarks about his own condition, or interactions with other people:

> "[On the Wabash River] from Mt. Carmel to the mouth of the Wabash, distance 125 miles, the voyager has but two places to fear. These are the Grand and Little Chains, which are beds of rocks, and in low water are quite dangerous, as the river runs over them with fearful rapidity. The Grand Chain just now presents quite a lively activity. There are now at work cutting a channel through some seventy-five men and half as many teams, hauling and making dykes. The channel when completed...will enable boats to go through in low water. Already 1300 feet of this is completed. The contractors hope to have the work completed this fall, if high water does not prevent. It is the opinion of the old settlers along the river that the Wabash can not be made navigable in low water.
>
> Two days running will bring the voyager to the Ohio, where he will meet with other crafts, from the homely flatboat to magnificent steamers; the emigrant with his household goods (which generally consists of ten or twelve children); men in skiffs all voyaging South, and one will be asked a dozen times a day, "Where is you a going? Is ye a going to Arkansas? How's the crops in Indianer?" The tide of emigration seems to be going to Arkansas.[11]

After eleven weeks, Ernest Morris and *Lucy* reached New Orleans, having covered over 1000 miles. That period included one week where a fever kept him idle--another indication that his constitution was not the strongest. He also arrived with no remaining funds, and was forced to find work as a hotel clerk.

A couple of weeks after his arrival in New Orleans, Morris wrote a description of the city for the readers of the *Indianapolis Sentinel,* focusing on the city's famous landmarks: the French Market; Jackson Square, with its statue of Andrew Jackson; Canal Street; and the above-ground vaults of the local cemeteries. Morris does not mention that the city was tense from recent unrest, but alluded to it indirectly:

> The first place that most sightseers seek is the levee, which extends along the river for seven miles. Vessels of every size and description, and from every nation, are receiving and discharging their cargoes; steamboats, large and small, ploughing their way up and down the river. Others discharging their immense loads of cotton, sugar, and molasses, together with the shouts of the Negroes, make a scene never to be forgotten. In the middle of the

> river are anchored four United States war vessels, and also the British Man of War which was sent here to protect her Majesty's objects in case of riot during election.[12]

Morris had disembarked in New Orleans two months after the most violent incident of the Reconstruction period, the Battle of Liberty Palace. On September 14, 1874, an armed insurrection of five thousand members of the Crescent City White League, composed mainly of former Confederate soldiers, forcibly removed the Republican governor, took over the statehouse and armory, and overwhelmed the city police and state militia. President Ulysses S. Grant quelled the rebellion by quickly sending federal troops to the city. The troops were still there when Ernest Morris arrived--and the recalcitrant Democrats were still not acknowledging the Republican elected officials. As years went by, the Battle of Liberty Palace became a symbol of racism for disenfranchised black residents--and a memorial for white supremacists.

Ernest Morris, young and apolitical, made no note of the tense atmosphere; nor did he remark on the social conditions in the South that former slaves endured, though the evidence was everywhere. Morris appeared to start out with the belief that the white Protestant culture in which he was raised was more civilized, productive, and moral than what existed elsewhere. At eighteen years of age and instilled with Indiana Presbyterian values, was not impressed by the religiosity of New Orleans, or rather the lack thereof. He concluded:

> A person coming from the North would be shocked at the way the Sabbath is observed. Here Sunday is the principal day at the theaters, markets, and for baseball matches and chicken fights. Opposite one of the largest churches in the city, baseball is played all day; saloons, stores, etc. open--nothing to indicate that it is Sunday. Verily, New Orleans is demoralized.[13]

Morris spent about six weeks in New Orleans, garnering the money needed for the next phase of his journey. He likely realized that *Lucy* was ill-suited for a journey along the Gulf coast, and instead disposed of his gallant canoe and boarded a ship bound for Jacksonville, Florida. In the 1870s, Florida still lacked a statewide railroad system, so the pathways to the settlement of the interior of the state were waterways--primarily the meandering St. Johns River and its tributaries. Without his canoe, Morris was compelled to complete his specimen collecting journey by taking small steam launches up the St. Johns. Moreover, he had arrived in Florida in winter, when Jacksonville was the primary destination for Northern tourists looking to escape the cold; therefore all boarding, food, and dry goods costs were high. The two letters that Morris sent to the *Indianapolis Sentinel* were full of complaints about the high price of everything (except fruit.)

Morris opted to camp outdoors, but even so, he wrote: "In closing I would say, whether to the tourist or sportsman, buy everything needed at home, if you come to rough it, which in my opinion is the best way. Bring your camping outfit with you, for the articles necessary for camp life are very high."[14] He also found it difficult to collect birds in the interior without his own boat, and was vexed by mosquitoes and sand flies. His stay on the St. Johns was not long--instead he opted to stay along the coast near St. Augustine. Between late January and March 1875, he shot birds and collected sea shells, starfish, and coral. Yet his letters for the newspaper hardly mentioned flora and fauna--instead he offered tips for sightseers and prospective homesteaders. Later, in South America, he would become a more skilled nature writer.

In Florida, Ernest also repeated his low opinion of the Southern work ethic:

> If it were not for the North and Northerners what would the people of Florida do? They depend upon the North for nearly everything. They raise hardly anything, they have no schools, nor do they seem to care. Travel Florida over and you will see the same general carelessness and shiftlessness. Their farms are poor and badly managed, and yet Florida, her lands, etc., have been compared to the West. Where in the West will you not find fine farms, and prosperous schools and communities? Energy and enterprise has done this. Energy and enterprise are what the people of Florida want to make it one of the greatest fruit-growing states in the Union.[15]

Morris returned to Indiana with a specimen collection that he sold for about $200, most of which came through displays in M. H. Spades dry goods store in Indianapolis. He had intended from the start to use this as seed money for an even more ambitious journey: a collecting trip to South America.[16] He told the *Indianapolis Sentinel* that he planned a two-year expedition, first sailing to Aspinwall, Panama; crossing the isthmus, and then taking a ship from Panama's Pacific coast to El Callao, Peru, near Lima. From El Callao, Morris intended to go inland to the headwaters of the Amazon, and traverse its entire course via a small boat.[17] His plan was admirable in scope and foolhardy in planning. It had all the earmarks of a swift and tragic undertaking.

---

[1] Boone County, Indiana Genealogy Trails History Group. "Boone County Indiana's First Judge." http://genealogytrails.com/ind/boone/firstjudge.html (accessed January 29, 2017)
[2] Ancestry.com. 2006. *Public Member trees* [online database]. Provo, UT: Ancestry.com Operations. Entry for Ernest Turner Morris, 1856-1891.
[3] "A Christian's Life Work," *Thornton* (IN) *Argus*, January 1, 1898.
[4] "Among the School Marms," *Indianapolis Sentinel*, January 31, 1874.
[5] "Funeral Notice," *Indianapolis Journal*, April 15, 1871.
[6] "Court Record," *Indianapolis Sentinel*, December 19, 1873.
[7] "Home Notes," *Indianapolis Sentinel*, April 20, 1874.
[8] "Afloat," Indianapolis Sentinel, August 26, 1874.
[9] "City News," *Indianapolis News*, August 26, 1874, p. 3.
[10] "Down the River," *Indianapolis Sentinel*, October 4, 1874.
[11] Ibid.
[12] "Down to the Delta," *Indianapolis Sentinel*, December 15, 1874.
[13] Ibid.
[14] "Correspondence–Florida," *Indianapolis Sentinel*, January 20, 1875.
[15] "Correspondence–American Antiquity," *Indianapolis Sentinel*, March 21, 1875.
[16] "A Hoosier Hero," *Milwaukee Sentinel*, May 5, 1875 p.5.
[17] "The Boy Naturalist," *Indianapolis Sentinel*, May 21, 1875.

3 THE NATURALISTS' PARADISE

While planning the route for his first voyage to the Amazon valley, Ernest Morris ignored the east-to-west course taken by many English and American scientists who had published books on the Amazon. Instead, Morris's preferred route was to follow a fictional trail described by a popular writer of boys' adventure stories: "Captain" Thomas Mayne Reid.[1] Reid's novels, targeted at English and American boys, were set in the wild places of the world, and were heavily laced with descriptions of the flora and fauna of those regions. In 1855 he penned a book titled *The Forest Exiles, or, The Perils of a Peruvian Family Amid the Wilds of the Amazon,* in which a family of Peruvian political refugees travels from Cuzco, Peru, across the Andes, and down one of the source tributaries (Madre de Dios) to the Amazon. Reid himself had never ventured to the Southern hemisphere, let alone the Amazon interior.

Young Ernest Morris dreamt of following in the footsteps of Reid's Peruvian refugees—without realizing that Reid made the feat of crossing the Andes and descending the Amazon's source streams much easier than it was.[2] The only concession that Morris made to the daunting prospect was to recruit a local friend to accompany him, a youth named Rodolphus S. Wilkin. Wilkin was the same age as Morris (nineteen in 1875), and had been born and raised in Indianapolis.

Once the young men arrived in New York in late May, 1875, someone--perhaps the Brooklyn collector-patrons they found[3]--dissuaded them from their plans to sail to the port of El Callao, near Lima, Peru. Through the mid-1870s, Peru had experienced great civil unrest, as followers of former Finance Minister Nicolás de Piérola battled the government of Mariano Ignacio Prado. Morris and Wilkin were advised that travel in Peru would be unsafe due to this civil unrest. Therefore, they decided to approach the Amazon from the conventional, safer approach: past the mouth of the river on the eastern coast of Brazil to the city of Pará (now known as Belém) on Marajó bay; and from Pará via steamer through the Breves channel onto the main river.

Morris and Wilkin embarked from New York City on May 29, 1875, on the schooner *Mary E. Douglass*, which made runs to Pará carrying mail, a handful of passengers, and light cargo. The voyage went swiftly, lasting just twenty-two days, but the last half of the trip must have been harrowing for the two young men: somewhere in the Caribbean, many of the ship's crew came down with "yellow jack," i.e. yellow fever, a deadly viral disease transmitted by mosquito. Before reaching Pará, the *Mary E. Douglass* lost her captain, first mate, the ship steward, and one other passenger.[4] Ernest Morris's first letters home from South America neglected to mention this; it was not until returning to the United States that Morris revealed the fearful circumstances of that first passage[5]. Once they arrived in Pará, Morris and Wilkin must have realized that they would be no safer, for the Amazon valley had frequent outbreaks of malaria and yellow fever. The cause and carrier of these diseases had not yet been identified, so the only precaution that one could heed was to avoid "sick" locales, defined as places with a bad miasma, i.e. an oppressive and unhealthy atmosphere.

The two young Americans spent a week in Pará, where they picked up advice on outfitting themselves for an extended stay in villages upriver on the Amazon. Pará was then a city of about 25,000 people, including a small community of English-speakers. Neither Morris or Wilkin knew any Portuguese, and were at a severe handicap once they left Pará. Fortunately for them, Ernest Morris learned languages quickly, and within a matter of weeks found himself able to converse not only in basic Portuguese, but in *língua geral,* the simplified version of the native Tupi language that had been promulgated during Brazil's colonial era.[6]

Morris and Wilkin had one distinct advantage over other Amazon travelers who made their journeys prior to the mid-1850s. Before then, travel up the Amazon from Pará was made by sail or human-paddled vessels, since steamers had not yet been put into regular service. Often, getting to the main channel of the Amazon from the Pará river involved a slow journey through connecting channels, sometimes requiring towing of the boats using ropes and manpower. Morris and Wilkin boarded a steamer and arrived at Santarem within a couple of weeks of leaving Pará—a trip that had taken months for the naturalists Bates, Wallace, and Spruce in the late 1840s.

Breves Channel.

Breves Channel by James Wells Champney. From: Smith, Herbert H. *Brazil, The Amazons and the Coast.* New York: Charles Scribner & Sons, 1879. [Public Domain]

From Pará, Morris and Wilkin took a steamer up the Pará river to Breves on the Rio de Breves, which connects to the Amazon via a series of channels. From there the steamer plied its way up the Amazon to the town of Santarem, roughly 900 miles distant from Pará. Santarem, with a population of 2500, was the largest settlement to be found along the entire span westward from Pará.[7] The only larger town along the whole length of the Amazon was Manaus, located another 500 miles above Santarem. Santarem is situated on the Amazon near the mouth of one of its main tributaries, the Tapajós River, and was an important trade center.

Morris might have imagined himself to be following in the footsteps of other famous naturalists—while Wilkin appears to have just been along for the adventure. How much of the history of the scientific exploration of the Amazon Morris had studied is questionable. In Morris's later writings, he mentions a few American and English explorers of the Amazon who wrote popular accounts, but made no reference to the published works of famous German, French, and Brazilian naturalists. A short review of the history of the notable nineteenth century scientific expeditions to the Amazon might be helpful in order to set Ernest Morris's depth of knowledge in context.

Until the nineteenth century, the Portuguese colonial administrators of Brazil discouraged any foreign travelers from journeying up the Amazon. However, to begin a summary of scientific exploration of the Amazon and limit it to Brazil would ignore the importance of one of the giant figures in natural science, Alexander von Humboldt. Humboldt made journeys up the Orinoco River in Venezuela in 1800, reaching it as far as the Casiquiare canal that united the Orinoco to the waters of the Amazon. A few years later, Humboldt made a separate journey to the Andes mountains of Ecuador and Peru, reaching the headwaters of the Amazon in those areas. These two journeys allowed Humboldt to describe a vast number of new species, many of which were later found on the Amazon.

As the political winds started to change on the eve of Brazil's independence, among the first to venture up the Amazon in 1819-1820 were the German naturalists, Johann Baptist von Spix and Carl F. P. von Martius. Martius concentrated on cataloging plant species and Spix on collecting zoological specimens. Both published several volumes illustrating their discoveries, but only one work by Spix was translated into English. Still, their work was very influential, and stands as the foundation for scientific studies of the lower Amazon.

Spix and Martius were hosted in Brazil by a German-Russian diplomat/naturalist, Baron von Langsdorff. Langsdorff mounted his own expedition of discovery in 1826-1828, plotting a dangerous course from Cuiabá north to the Rio Arinos, and following it to the upper Tapajós. From there, Langsdorff descended the cascades of the Tapajós to the Amazon, but the journey was so arduous that he went insane from exertion. However, the expedition did reach Santarem at the mouth of the Tapajós successfully. Ernest Morris would later believe that he was the first Western naturalist to see the Upper Tapajós, but he was obviously unaware of Langsdorff.

At about the same time as Langsdorff, a French naturalist named Alcide Charles Victor d'Orbigny undertook a grand tour of South American countries, including the Amazon and Madeira Rivers, and returned to France with over 10,000 specimens. Charles Darwin corresponded with d'Orbigny and hailed his accomplishments. A German botanist, Eduard Friedrich Poeppig, travelled through Chile, Peru, and the upper reaches of the Amazon in between 1827 and 1832. He brought nearly 14,000 specimens, mostly botanical, back to Germany. Neither d'Orbigny or Poeppig's works were translated into English, but, like Spix and Martius, throughout the nineteenth century their findings and illustrations were disseminated throughout the scientific community.

Explorations on the lower Amazon were interrupted between the late 1820s and through the 1830s by the Cabanagem movement, a social revolt by northeastern Brazil's natives, freed slaves, and *mestiços* against the white landowners and merchants of the state of Pará. Despite the political and physical dangers, in 1834-1835 two British naval officers, William Smyth and Frederick Lowe, sought a navigable water passage from Lima, Peru across the Andes and down the entire length of the Amazon to facilitate a west-to-east trade route. Their work inspired numerous future expeditions to expand this cross-continental concept, which came to include the idea of a railroad along the Madeira River--a venture that would eventually recruit Ernest Morris.

Another explorer, Robert Hermann Schomburgk, a German-born businessman, was entrusted by Britain's Royal Geographical Society to survey British Guiana in the years 1835-1839. His journeys led him to the northern section of the Amazon valley, including parts of Venezuela and Brazil. Schomburgk discovered many new orchid varieties—so many that an entire genus of plants was named in his honor—as well as the giant water lily, Victoria regia. Many of his accounts of explorations in Guiana were first published in English. Schomburgk also translated the account of Prussian Prince Adalbert's 1842-1843 voyage up the lower Amazon and Xingu Rivers.

Prince Adalbert was not the only traveler to take advantage of the cessation of the Cabanagem insurrection in state of Pará. Louis-Marie-Fronçois Tardy de Montravel was commissioned by the French Navy to conduct a survey of the coast of Brazil, French Guiana, and the lower Amazon. Montravel compiled detailed maps of the area during the years 1842-1845. To the west, François de Laporte de Castelnau was sent by French King Louis Philippe I from Rio de Janeiro to Lima, Peru; and from there turned eastward across the Andes and down the Amazon. Castelnau produced several volumes of finely illustrated botanical and zoological specimens, often reproduced in later publications throughout the world.

In 1846, American businessman and amateur entomologist William Henry Edwards journeyed up the Amazon, compiling a collection of butterflies and marveling at the variety of flora and fauna. He then wrote *A voyage up the river Amazon including a residence at Pará* published by Appleton in 1847. Edwards' enthusiasm for the Amazon as a naturalist's laboratory greatly influenced Henry Walter Bates and Alfred Russel Wallace, two young English naturalists who decided to journey to Brazil in 1848. Bates and Wallace were later joined by botanist Richard Spruce, and together the three men have been hailed as the foremost figures of the scientific study of the Amazon, describing over 8,000 new species. Wallace remained in Brazil until 1852, while Bates stayed eleven years, until 1859. Bates then penned the foremost travel narrative of Brazil by a scientist, titled *The Naturalist on the River Amazons.*

Before arriving in Brazil, Alfred Russel Wallace had read and supported the concept of the transmutation of species, i.e. the alteration of one species to create another. This theory, which was not widely accepted, was a forerunner to the idea of natural selection as the mechanism for evolutionary change. While in Brazil, Wallace became intrigued by biogeography, particularly the proximity of one species to another closely-related species in relation to geographical barriers. As an example, Wallace wrote about the differences in species (like monkeys) on the north and south banks of the Amazon. After returning to England and later embarking on another journey to Borneo, Wallace began a correspondence with Charles Darwin. That exchange of ideas resulted in the joint reading of essays that both men wrote about natural selection; followed later by Darwin's 1859 landmark book, *On the Origin of Species.*

During the same years that the three great British naturalists—Bates, Wallace, and Spruce--were traveling through the Amazon valley, two American naval officers, William Lewis Herndon and Lardner Gibbon, were charged with conducting a commercial survey of the western Amazon valley, noting natural resources and transportation potential. Their findings, published by the U. S. Navy as an extensive report, *Exploration of the Valley of the Amazon,* was widely circulated and encouraged many entrepreneurs to investigate the economic possibilities of the Amazon region. In the early 1860s, Englishman William Chandless made a similar survey of the Purus River and other southern tributaries, including the

Tapajós, for the Royal Geographical Society. Again, Ernest Morris appeared to have been unaware of Chandless's published account, likely because it was published in the minutes of the Royal Geographical Society, rather than by a commercial publishing house.

Brazilian João Martins da Silva Coutinho made several expeditions in the 1860s to collect geological specimens for the Museu Nacional and made investigations into the rubber industry that would have dramatic repercussions. Rubber harvesting had already transformed the lives of the indigenous peoples for the worse; Silva Coutinho both proved the superiority of Brazil's rubber trees and recommended a more efficient, scientific cultivation of the trees on plantations, whereas in nature they grew far apart from one another in mixed forests. His efforts to promote rubber tree horticulture were noted in Europe, eventually resulting in rubber plant seeds being smuggled out of Brazil for cultivation in India and Southeast Asia—in effect causing the eventual collapse of the industry in Brazil. Silva Coutinho also accompanied American biologist Louis Agassiz on the Thayer Expedition on the Amazon in 1865-1866.

The Thayer Expedition was another hallmark event in the timeline of exploration of the Amazon. It was led by America's leading biologist and geologist, Louis Agassiz, accompanied by his wife and a team of a dozen scientists headed by Cornell's Charles F. Hartt. Their primary focus was on the fish and the geology of the region, but Agassiz had an ulterior motive of studying South American Indian characteristics in order to further his polygenism views on the origins of race. Polygenism is the theory that mankind originated in separate areas of the globe, giving rise to different races. In the eighteenth and nineteenth centuries, many believed that polygenism was more scientific than monogenism, since the religious account or man's origins in Genesis was an example of monogenism. Agassiz believed that a Creator had made different ecological zones populated with unique species and races of men, each with their own characteristics.

Though not a defender of slavery, Agassiz's theories were embraced by those who believed that racial differences implied superiority and inferiority. In contrast, Darwin's second book, the *Descent of Man,* extended the concept of evolution to humans, and reestablished monogenism on firm scientific ground. Although Agassiz's misguided polygenism opinions did not stand the test of time, his presence in Brazil did encourage the founding of the Goeldi Museum of Natural History in Pará by Domingos Soares Ferreira Penna. Moreover, Agassiz's wife

Elizabeth wrote an engaging popular account of the expedition, *A Journey in Brazil.*

In 1867, the Empire of Brazil opened navigation on the Amazon to vessels of all nations, with the aim of encouraging more trade with neighboring countries. To this end, German engineer Franz Keller-Leuzinger was commissioned by the Brazilian government to survey the possibility of routing a railroad along the Madeira river to avoid the waterfalls that made navigation by water impossible. The same year, American James Orton ventured from Quito, Ecuador across the Andes, descending the Napo River to the Amazon. Orton's journey was written up in a widely-circulated book, *The Andes and the Amazon,* published by Harper in 1870. Orton made two further trips up the Amazon and Madeira rivers from east to west in 1873 and 1876, dying on Lake Titicaca near the end of the latter journey.

Prior to Morris's arrival in Brazil, the most prominent Amazon journeys of the 1870s were the Morgan Expeditions led by Charles F. Hartt, accompanied by his students from Cornell. Those separate annual trips, starting in 1870, examined a range of the sciences, from flora and fauna to geology, paleontology, and ethnology. One of Hartt's assistants on the 1870 expedition was a young man named Herbert Huntington Smith, a geologist by training, who developed interest in the other sciences while with the expedition. Smith returned to the United States to complete his degree in 1872, then returned to Brazil on his own in 1874.

The length of the timeline of exploration listed above should make it clear that by the time Morris and Wilkin arrived in Santarem in July, 1875, an enormous amount of survey and scientific work had already been conducted in Brazil. Almost all the studies had been made from areas near settlements on the great river's waterways, the only accessible parts of the interior. Bates himself had remained for many years in the town of Santarem, which had a pleasant climate and was close to several different habitats: rivers, lakes, and small streams; beaches; várzea floodplains and islands; sandy campo plains; and thick rainforest.

Ernest Morris might have been knowledgeable for his age, a skilled outdoorsman and a good taxidermist; but he lacked many of the other skills required of a scientific observer: map drawing; scientific illustration; use of instruments of measurement; chemical analysis; etc. He travelled without any of the tools needed for these tasks. Morris, on this first trip and on others, made no claim that he was engaged in scientific expeditions of discovery. Instead, he saw himself as a collector, whose aim was to sell and deliver specimens to scientists and hobbyists in the United States. He might have hoped to come across something entirely new to science, but probably realized that the chance of doing so was low.

As for Rodolphus Wilkin, his degree of interest in natural history may have been nonexistent. After the fact, Wilkin told one newspaper that he went to Brazil "for fun."[8] Once situated in Santarem, it is unclear how often Wilkin accompanied Morris on his collecting trips; Morris stated that his later forays were made alone, having been forced to leave Wilkin in Santarem suffering from some tropical disease. Wilkin published no letters and gave no interviews after the two returned to Indianapolis, though one paper remarked that he had no further travel plans and had "abundantly satisfied the cravings of his nature for adventure."[9] He later went to work for the Atchison, Topeka, and Santa Fe Railroad, and never evidenced further interest in collecting.[10]

When the two young Americans reached the village of Santarem, they encountered a lively town of 2500 people, a blend of white foreigners, Portuguese descendants, South American Indians, blacks (free and enslaved), and persons of mixed-race. The wealthier residents had mud-walled houses, while the poorer, working section of town, the *aldeia*, consisted of thatched huts. The two most substantial buildings of the town were the cathedral and the jailhouse. Due to the extended shallows of the shoreline, there were no docks; steamers anchored at a distance and all passengers and cargo were transferred using small canoes and boats.

Immediately on arrival, Morris observed the main industry of Santarem, piraracu fishing:

> Fishing forms one of the occupations of the Indian, the largest fish being the piraracu (*Sudis gigas*) the name signifying in the língua geral the ancient language of the Indians, "red fish," so called from the color of the fish, although there are two species, one of which has pure white flesh; the scales also, which are very large, are deeply margined with red. This fish abounds in the Amazon, in all the tributaries below the *cachoeiras* (rapids) and in the great lakes. It attains an immense size and feeds upon the *capim* or water grass; as it will not take a bait it is always captured by spearing; the fishing being generally conducted at night. The fish is simply salted and dried in the sun; it is eaten boiled or broiled and in taste resembles dried cod fish but is much nicer;

when boiled fresh it somewhat resembles halibut, and a piece of fresh piraruc u roasted on a stick makes a dinner which any epicure might be glad to eat. The chief local trade between the Amazonian ports is in dried pirarucu, and immense quantities are annually shipped from Santarem to Pará. The tongue of this fish, which is very hard and rough, is used as a grater; and the Indian women wear the scales as charms.[11]

The Pira-Rucu (*Sudis gigas*) by Franz Keller. From: Keller, Franz. *The Amazon and Madeira rivers; sketches and descriptions from the note-book of an explorer.* Philadelphia: J. B. Lippincott & Co., 1875. [Public Domain]

Despite its small size, Santarem of the late 1870s was a waystation for people with wildly varying ambitions. Six American families lived on scattered ranches near Santarem, the surviving members of a colony of about 140 Confederate refugees that fled the United States following the Civil War. The presence of these expatriates had been mentioned by Agassiz and Hartt; and had been written about in newspapers in the United States. Most of the original 140 refugees did not have the skills or wherewithal to thrive in Brazil. They earned a reputation for drinking and other disagreeable behavior, and soon gave up and left the country. However, a few adapted, and managed successful sugar cane and cattle farms for over a decade, building some of the finest houses found anywhere on the Amazon.

There was also a group of English expatriates, consisting of the extended family of an inept plantation owner and adventurer named Henry Wickham. In the early 1870s, Wickham had been recruited to procure *Hevea brasiliensis* seeds for the Royal Botanical Gardens at Kew, a project overseen by Clements Markham, who had earlier spearheaded a similar effort to procure seeds of the cinchona tree, the source of quinine. This was hardly an academic exercise: *Hevea brasiliensis* was the world's best quality source of rubber latex, and rubber was one of the main resources powering the industrial age. Great Britain did not want its economic future tied to a natural resource monopolized by another country, and so it had an imperial interest in obtaining those seeds and transplanting them to one of its own tropical colonies in India or Southeast Asia. By 1875, the relatives that had accompanied Wickham had given up on plantation life and instead operated an English school in Santarem to educate the children of the *Confederados*. Henry Wickham and his wife Violet remained on their poorly-maintained piece of land in the highlands south of Santarem, near the Rio Curua-Una. Previous attempts by Wickham to procure rubber seeds for the botanists at Kew had met many difficulties, but he made another more successful effort during the same months that Morris and Wilkin resided in Santarem.

It probably did not take long before Ernest Morris heard that a guest at one of the Confederado plantations was another young American naturalist, Herbert Huntington Smith—who had accompanied Hartt on the 1870 Morgan Expedition. There is no documentation that Ernest Morris and Smith met—but the fact that they were both located in the Santarem vicinity for many months of 1875-1876, if not a full year—and were engaged in the same activity of collecting flora and fauna specimens, makes it a near certainty that each knew about the presence of the other. Moreover, the fact that neither man mentioned the other, combined with the timing of their subsequent writing efforts, makes it likely that they viewed each other as unfriendly rivals.

Herbert Huntington Smith had many advantages over Ernest Morris. In 1875, he was twenty-four years old, five years older than Morris. He was a Cornell graduate and a trained geologist, having been a member of Charles F. Hartt's 1870 Morgan Expedition to Brazil.[12] He was also a talented writer, having served for a time as an assistant editor for the *Fayetteville* (New York) *Weekly Recorder*.[13] Though his father had died when he was just a child, Smith was raised with wealth and household servants. His grandfather had been a founding father of Manlius, New

York; a state legislator; and a trustee of several colleges. His great-grandfather fought in the American Revolution.

Following his studies at Cornell, Herbert H. Smith arranged the sale of his family's home in Manlius, realizing enough capital to fund an extended, independent stay in Brazil.[14] He returned to the Amazon in 1874 to collect specimens. Once in Santarem, he renewed an acquaintance with Romulus John Rhome, one of the ex-Confederate officers who had resettled near Santarem at a place called Taperinha, about twenty miles east of Santarem on a bluff overlooking the Amazon. It is likely that Smith, during the years 1875 into early 1876, resided at Taperinha or in the relative comfort of another Confederado plantation house. Morris and Wilkin, on the other hand, built their own low-budget hut on the outskirts of Santarem and suffered the numerous discomforts of sharing their domicile with the pests of the rainforest: ants, scorpions, vampire bats, cockroaches, and tarantulas.

Taperinha Plantation, from the River.

Taperinha Plantation, From the River by James Wells Champney. From: Smith, Herbert H. *Brazil, The Amazons and the Coast.* New York: Charles Scribner & Sons, 1879. [Public Domain]

Left to their own devices—with no help or advice from Herbert H. Smith—Morris and Wilkin began their collecting efforts at an inlet just east of Santarem called Maicá. Morris later wrote:

> About five miles east of Santarem is a small river or rather inlet, called by the Indians Maicá[15]; about a mile from the mouth is a little Indian settlement composed of small mud palm-thatched houses scattered along the banks of the river...Our excursions to

> Maicá were among the pleasantest from Santarem; we were always most hospitable received, and many hours were most pleasantly passed in eating oranges and watching the Indian girls making farinah, which with *pacovas* and fish forms the food of the people. Although communication is easy by canoe, our trips were usually made by land, as thus we were better able to study the natural history of the country. Starting early in the morning in order to avoid the intense heat of the sandy campo over which the road, which is good, runs for the first three miles, we reach the wooded country. Barren as the campo is, it is not without some feature of interest; the low trees owing to the prevalence of the easterly trade winds all lean towards the west; some of the trees are at seasons bright with flowers. The cashew trees are bright with yellow and red fruit, peculiar from having the seed, which resembles a large kidney bean, projecting from the apex, and flocks of gay paroquets which build their nests in the hollows of the trees fill the air with their ceaseless chatter.
>
> Leaving the campo, the road winds round the base of a high hill from the top of which a magnificent view is obtained embracing the blue Tapajós, the far reaching Amazon and even the hills of Monte Alegre some forty miles away, which in this clear atmosphere are distinctly visible. We soon reach the denser forest, the huge trees interlace their branches high overhead, it is delightfully cool, the air is full of the perfume of innumerable flowers and many gay birds and brilliant butterflies flit past...Thus we go from house to house, collecting many rare and beautiful butterflies and beetles in the wood paths, and are everywhere received with similar hospitality, returning to the city late in the evening.

It was on the walk to Maicá that Morris was first exposed to one of the great scourges of the Amazon basin, the sauba ant. The sauba is a species of leafcutter ants, which cut live foliage and return it to their nests to decompose—the sauba hive lives on fungus produced by the rotting biomass. Morris was equally fascinated by the human use of the sauba:

> In the month of June, we often came upon groups of Indian children sitting patiently in the broiling sun and catching sauba (*OEcodoma cephalotes*[16]) queens. This ant attains almost an inch in length and throws up nests many feet in diameter; and is so destructive to vegetation. This queen ant is solitary in its

habits and lives in small holes in the sand. Each child has a long straw and a *cuia*[17] (made from the fruit of the calabash tree) full of water; he thrusts this straw into the hole of the ant, the insect seizes it and is quickly drawn out and thrown into the water. The body, which is very large, is parched and eaten; it has a strong, spicy taste resembling cloves and is not unpalatable. They are also eaten as sauce mixed with a preparation of mandioca[18] forming a spicy yellowish paste.

Sauba or leaf-carrying ant by Edward William Robinson. From: Bates, Henry Walter. *The Naturalist on the River Amazons.* London: John Murray, 1863. [Public Domain]

Also along the trail to Maicá that Morris first observed the process of making farinha, the food staple that was to sustain him on all his travels in South America:

> This is prepared from the root of the mandioca or cassava (*Manihot utilissima*[19]), a euphorbiaceous plant with long-stalked palmate leaves and small dioecious flowers at the ends of the branches. At the beginning of the rainy season in December, the old stalks are cut into lengths of about three inches and are planted four on a hill, the lower end being inserted obliquely into the ground; these cuttings immediately take root and begin to grow and in a few months the plants attain a height of five feet, but the roots are not usually fit for use for nine months and are frequently not dug until they attain a large size and are many years

old. There are two varieties, one with a yellowish the other with a white root, but both are indiscriminately used in making farinah. The rough skin is first scraped from the large fleshy roots with a long knife, the roots in the preparation of one kind of farinah having been soaked in water for several days.

They are then reduced to a pulp by grating, and the mass is then put into a *tapiti*[20], a long wicker work tube with a broad loop at the lower end, a long lever is inserted, upon the end of which two or three women sit: the tube, which is about nine inches in diameter and very flexible, contracts and the juice runs out oat the lower end, so that as the pressure is great the pulp is soon almost dry. The mass is then pounded in a large hard wood mortar and then sifted to remove any root fibre or coarser portions. Every Brazilian house has under an adjoining shed a large earthen fire-place, the whole top of which is covered by an earthen pan about six inches deep. This is called a *forno*. Into this, when well heated, the meal is put and constantly stirred from four to six hours according to the intensity of the fire. The steam first arising is very disagreeable but is soon dissipated and the meal has a sweet, pleasant smell. When done it is of a light yellow color and somewhat resembles bread crumbs. It is eaten without further cooking or is drank mixed with water. To a stranger is seems hard, tasteless, totally unpalatable and only fit for food for hens; it has great absorptive properties, is very filling but does not possess much strengthening nutriment.

It is on this that the Indians and the lower classes of Brazil chiefly live, they regard it as an essential, whatever else they may have. We once had a mutiny among our boat crew because we had neglected to provide farinah although we were well supplied with fresh white bread. On entering an Indian village one at once notices the protruding stomachs of all the children, they seem to be all far gone in the period of maternity; this unnatural distention is produced by their living on dry farinah.

The expressed juice of the mandioca, if allowed to settle, deposits a large portion of starch, from which by simply heating it, is prepared the tapioca of commerce, the starch grains from the heat swelling and bursting and becoming agglutinated together. The juice expressed in the rainy season deposits more tapioca, while the dry season is preferred for making farinah. There are

> also many drinks made from the mandioca, some of which are slightly acid and refreshing and when fermented intoxicating; the mode of making some is such that no stranger who has witnessed the preparation cares to partake of the beverage. A strong *cachaça* also distilled from the root, and the only starch used in the Amazonian valley is tapioca.[21]

In the late summer of 1875, Morris and Wilkin embarked on a lengthier journey, venturing thirty-five miles south of Santarem to the forests near the Rio Curua-Una—the same vicinity where the restless rubber-tree thief Henry Wickham was trying to maintain a plantation. After returning to Santarem, Wilkin came down with an illness—likely malaria—and left further expeditions to Morris alone.

In January, 1876, Morris hired a 20-foot canoe, called a *montaria,* and with two men and a boy left Santarem and traveled westward on the Amazon river past the mouth of the Tapajós, with the intent of making an extended tour of a huge lake, Lago Grande do Villa Franca.[22] The lands surrounding the lake supported cattle ranches and cacao plantations. Morris was invited by one prosperous cattle rancher to make an extended stay at his house. He accepted, and during his stay observed the workings of a nearby cacao operation:

> The cacao (*Theobroma cacao*) is largely grown on the middle Amazon. It is a small tree or rather large bush, seldom exceeding fifteen feet in height. It has large, oblong, taper-pointed leaves which when young are rosy and very beautiful. The flowers are small, fragrant, yellowish and are borne in clusters on the trunk and larger branches. The fruit is about ten inches long, vary however greatly in size, and about five inches broad; it is oblong, blunt and is marked with ten elevated ribs running lengthways, and is when ripe bright orange-yellow. The shell is very hard. Although authorities state that the fruit contains from fifty to 100 seeds we have never been able to find one containing fifty; the average is from thirty to forty-five. The seeds are embedded in a white semi-acid pulp; it is from the seeds that the cocoa of commerce is prepared.
>
> The fruit is when ripe gathered by hand and is broken transversely by a large knife. The mass of pulp and seed is taken out and is either left for a day to slightly ferment or is at once put into a *tapiti* and pressed almost dry. It is then dried for one to three days in the sun and is ready for market. The juice running

from the *tapiti* is called *vinho da cacao* (wine of cacao) and is a most refreshing and delicious drink. It is slightly acid, aromatic, and possesses a flavor peculiarly its own. Unfortunately, it soon ferments and can only be preserved in its freshness a single day; by boiling, however, fermentation is checked but all the delicate aroma of the wine is lost. It is almost worth a voyage across the Atlantic to drink the wine of cacao in its freshness. This drink is very healthy and one can consume an immense quantity. I one could only add to it a small piece of ice, a thing not procurable, it would be perfection. The word *vinho* (wine) does not in Brazil necessarily imply a fermented liquor, but every infusion of fruit in water, of which many are used as drinks, is so called.

A superior white vinegar is also prepared from the cacao and the juice is boiled down into a sauce which in taste and color much resembles Shaker apple sauce. From the *vinho da cacao* a strong and very rich flavored brandy is distilled. The husks contain a large amount of potash and are burnt to yield a lye which is used in making soap. Great quantities of the cacao fruit are annually destroyed by monkeys and parrots which are very fond of the seed. The cacao is planted from the seed and the plants yield fruit when three years old; they continue bearing for many years. We know of many orchards which are more than seventy-five years old. When first planted, lines of plantains or bananas are set out alternately with the rows of cacao, to protect the young plants from the sun. These bear fruit in one year and are cut away as soon as the cacao plants attain some size.

One who has not seen cannot imagine the deep gloom of a cacao plantation; the rows of trees are planted so close that the branches interlace overhead and the broad dense foliage make it almost dark even at noonday. The ground is thickly carpeted with the large dead leaves among which is the favorite haunt of one of the most poisonous of Brazilian snakes, the *jararaca*, (*Bothrops jararaca*) and no grass or plants of any kind are to be seen. When in fruit, however, the effect is very pretty, the trunks hanging thickly with the bright orange-colored fruit and the bright butterfly *Morpho Hecuba* and red and black *Helioconias* are by no means uncommon.[23]

It was during this tour of Lago Grande that Morris saw the enormous water lily, *Victoria regia*[24], for the first time.

While touring the Lago Grande region during the early months of 1876, Morris was unable to send or receive letters on the Amazon steamers, leaving him out of touch with his father back home. His last letter to the *Indianapolis Sentinel* had been printed in October, 1875—and described events up through July, 1875; and the last letter to his father was received in March, but was written just before Morris left for Lago Grande. Lacking any genuine news from Ernest Morris, some mischievous writer at the *Indianapolis Sentinel* decided to manufacture a fantastical version of his journeys:

> Morris Mesmerized: Romantic Story of a Soldier of Science—An Indian Princess Rescued from Death in a Tropical Storm, Provides for Her Heroic Preserver, and the King Unlocks His Treasures in Gratitude.
>
> If the ghost of a doubt lingers in this community that truth is indeed stranger than fiction, let it be at once dispelled by the following leaf of history from the romantic career of our young boy naturalist, Ernest Morris. The heroic adventurer, whose unconquerable zeal for science and its myriad beauties, has led him, like a second Agassiz, to penetrate the mysteries of nature, needs no introduction to this neighborhood, or to the readers of a paper which has from time to time published interesting scraps from his journal. How he shouldered his rifle, with no capital but hope and determination, sallied forth from Indianapolis nearly two years ago at the age of eighteen; how he made the voyage to New Orleans by river in a skiff; how he sailed in a South American merchantman to Pará, at the mouth of the Amazon, with the grand but apparently wild dream of pursuing his favorite study in the vast tropical forests which few have ever dared to penetrate; how he successfully accomplished the daring feat in the face of innumerable obstacles known as snakes and tigers; how he followed up the course of that mighty river in his search for the products of nature, sometimes on foot and again in his light Indian canoe till he had penetrated to the mountain springs of the Amazon far beyond the limits of Brazil in the territory of Ecuador; and how he has perseveringly enriched his collection with invaluable specimens of bird and insect life; all this, and more, in the way of side incidents, very romantic and thrilling, has been told in his own letters, in a tone of modesty which is the best index of his sincere devotion to science...

His father, Mr. Sanford Morris, has just received from his adventurous son a letter which reads like a leaf from the career of Sinbad...While the dashing naturalist was on his way back from the headwaters and had halted near Santarem one evening to take his customary lunch on monkey meat, leaving his light canoe sleeping quietly in a cove by the shore, he suddenly noticed by signs in the wind that a storm was about to bear down in great fury...far out in the stream, seated in a fragile canoe, was a young Indian girl, vainly struggling to reach shore before she drifted into the dangerous current...Ernest realized the peril at a glance and, without stopping to consider the question, plunged in to her rescue, and after a desperate struggle with the excited stream, pushing the canoe with one hand while he fought the current with the other, he managed to land his charge safely on the shore...His eyes were dazzled by her beauty; she surpassed any specimen he had met with yet, and he was particularly smitten with the girdle which bound her apron...it shone like fire, and well it might, for it was studded with enormous diamonds, more than enough to ransom any king of the Orient.

When the storm had subsided, Miss Indian insisted that her preserver should go home with her and be thanked by her papa. The youth had saved his only child, a blood royal Indian princess of the Amazon...Countless diamond, silver, and gold mines and treasures by the millions were his. In his gratitude the old warrior literally crushed Morris with his kindness...It was hard for a young man of twenty with nothing but a few insects in his purse to resist these dazzling splendors, and harder still to turn his back on the copper princess who loved him to distraction...[the father] finally proposed to him that he should return to America with his Indian beauty, introduce her to the strange people of his country, and present her to his parents. In case they were pleased with her beauty, accomplishments, and diamond dory, and thought she would make Ernest a dutiful and domestic little wife, that they were to return to Brazil and be married, and the young scientist should receive boundless wealth to prosecute his studies.

When the boy naturalist, the great diamond son-in-law prospect, comes back in June with his bewitching copper heiress, he intends to place the first book of his Amazonian journals in the hands of an eastern publisher, to lecture here and at various points in the East where he has already been invited to come for a

> consideration, and then "go home his crown to wear" in the heart of Brazil.[25]

This article, published by a respected newspaper, was, of course, a patchwork of lies concocted by some prankster to embarrass Morris. It is not difficult to conjecture who the fabulist was, since one of Ernest's old schoolmates was a young man named Barrett Paine—the same young man who had started out in the canoe with Ernest Morris on the White River in August of 1874. After their canoe tipped over early in the trip, Paine gave up and returned to Indianapolis, while Ernest continued onward and was universally praised for his determination. Paine had been publicly shamed for not demonstrating the same pluck as Morris. He was the son of Daniel L. Paine, a longtime Indiana newspaperman and noted poet, who had once worked for the *Sentinel* and later for the *Indianapolis News*. It is not difficult to assume that young Paine, who obviously had heard about or read a letter sent to Ernest's father about returning home, contrived a way to both embarrass Ernest Morris and the *Sentinel* by feeding them this fantasy about a diamond-clad princess.[26]

Unlike other news accounts about Ernest's trip to the Amazon, which had been limited to newspapers in Indiana, the *Sentinel*'s diamond princess article was reprinted in newspapers around the nation. Ernest's younger brother Frank was compelled to try to set the record straight. He wrote to the *Sentinel*'s competitor, the *Indianapolis Journal*:

> I noticed in the *Sentinel* of Tuesday a supposed communication from Ernest T. Morris. I wish to thank the gentleman (or fool) who composed it, and inform him that if he will keep in practice he will soon acquire the very pleasant notoriety of being the greatest liar of the day. Frank C. Morris.[27]

Once Ernest Morris returned to Santarem from his solo journey to Lago Grande do Villa Franca, he found that Rodolphus Wilkin was as sick as ever; and moreover, there were letters addressed to Ernest from Indiana stating that his father, Sanford Morris, was seriously ill. Without further delay, Morris and the ailing Wilkin took a steamer back to Pará in early May, 1876, and from there booked passage to New York.[28] At almost the same time, Herbert H. Smith also left Santarem to go to Rio de Janeiro to assist his mentor, Charles F. Hartt, conduct research for the Geological Survey of Brazil; and the elusive Henry Wickham also abandoned Santarem, taking with him 70,000 rubber tree seeds for the British Empire.

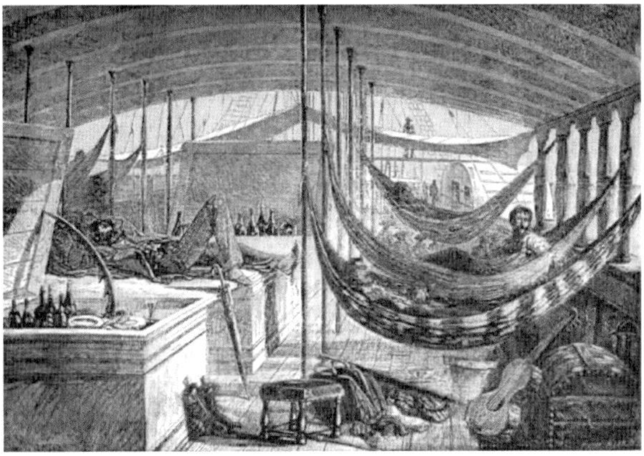

La dunette du vapeur *Le Marajo* by Édouard Riou. From: Biard, François. *Deux anness au Brésil.* Paris: Librairie de L. Hachette, 1862. [Public Domain]

Back in the United States, Morris disposed of the greater part of his collected specimens to Robert G. Eccles and John Akhurst, both of Brooklyn. Akhurst was a taxidermist by profession and a self-trained entomologist, a well-known figure both as a collector and as a supplier of insect specimens and collecting equipment. Akhurst donated much of his collection to the Long Island Historical Society's museum.[29]

Once Morris returned home to Indianapolis from Brooklyn, he rebuked the *Sentinel* for its "reportorial license," but did grant them an interview to set the record straight, which he concluded by telling the paper that in the future they should "go slow on the diamond princess."[30] There were no further reports that he was planning to get a book published—or even that he kept notes on his trip; but he did agree to conduct a few lectures on the Amazon at halls in and around Indianapolis. Perhaps he was stung by the cold shoulder he was given by the much more accomplished Herbert H. Smith, because the newspaper also made a point to announce that, during the summer, Morris would continue private studies on the natural sciences with Dr. Gilbert M. Levette, the Assistant State Geologist of Indiana; and in the fall, Morris planned a return to the Amazon.

At face value, it may seem odd that Morris worked with a geologist to further his studies in natural history, but Levette and others in State Geologist's offices were often called upon for their expertise in both the earth sciences and biology. Few universities offered a curriculum devoted to these sciences, but state governments needed geologists to map mineral resources, measure water tables, consult on public works projects, etc. Many state geologists like Levette were asked to help identify any strange object found in the ground, or any strange creature brought to their attention. Levette's office held displays of both animal and mineral specimens; and his shelves carried the best collection of flora and fauna identification and research materials found in the State.

Indiana's office of State Geologist also had an illustrious heritage. The first State Geologist was David Dale Owen, son of the social reformer Robert Owen. David Dale Owen conducted the first geological surveys of six states, and served as the State Geologist of Indiana, Arkansas, and Kentucky at separate times. He was succeeded in 1860 by his brother, Richard Dale Owen, a professor of geology at Indiana University. Following the Civil War, in which he served as a Lieutenant Colonel, Richard Owen agreed to serve as Purdue's first president. He conducted geological survey of four other states, and wrote widely on the subject, particularly on the idea of environmental determinism.

Richard Owen's position of State Geologist was next held by Gilbert Levette's superior, Edward Travers Cox. Cox, like the Owens, was a product of Indiana's utopian New Harmony community. Cox brought his large collection of minerals, fossils, shells, and other specimens to Indianapolis—collections which eventually formed the basis of the Indiana State Museum. Levette, as Cox's assistant, was an amateur geologist, mapmaker, inventor, and collector of natural history specimens. He did not discuss his early years with people in Indianapolis; but upon his death the *Cincinnati Enquirer* revealed that he had been married twice before arriving in Indianapolis. His second marriage in 1860 was to Emma J. James of Cincinnati. She gave birth to a son but herself died at an early age. Levette enlisted in the Union Army in 1863 as a private and mustered out as a surgeon. He located to Indianapolis after the war and soon was appointed as an assistant to Cox. It would not be an exaggeration to say that the Indiana State Geologist's office of the nineteenth century was the foremost State-sponsored scientific bureau in America. Ernest Morris, despite his impatience with academic work, likely learned a great deal from Levette.[31]

---

[1] Reid had an enormous influence on several generations of adventure readers, and was frequently translated in Eastern Europe and Russia. President Theodore

Roosevelt mentioned Reid as one of his inspirations as a young man, and adopted Reid's love of natural history. In the 1840s, Reid was a frequent drinking companion with Edgar Allen Poe, who called Reid, "a colossal but most picturesque liar. He fibs on a surprising scale but with the finish of an artist, and that is why I listen to him attentively."

[2] Reid's imagined route from Peru across the Andes was difficult, but not impossible, Reid may have known about the 1840s expedition of French naturalist Castelnau, or the 1851 expedition of American Naval officers Herndon and Gibbon.

[3] The named sponsors of Morris's first trip to the Amazon were Dr. Robert Gibson Eccles, a chemist and plant collector residing in Brooklyn; and John Akhurst, a Brooklyn taxidermist and entomologist.

[4] "Yellow Fever," *Cincinnati Daily Star*, September 8, 1875; "Yellow Fever," *Boston Post,* September 9, 1875.

[5] In his first letter home, Morris only mentions that the transit had been quick and that he and Wilkin had been subject to the customary hazing ritual performed on those who were crossing the Equator for the first-time, e.g. a trial by "King Neptune's Court." In Morris and Wilkins case, the hazing consisted of being woken at midnight, having their faces covered with paint, and then having the paint shaved off with wooden razors.

[6] "Correspondence—Valley of the Amazon," *Indianapolis Sentinel*, October 26, 1875.

[7] "City News, *Indianapolis News*, October 21, 1875.

[8] "The Young Naturalists," *Indianapolis Journal*, June 8, 1876.

[9] Ibid.

[10] In 1880, Wilkin was working for the Acheson, Topeka, and Sante Fe railroad when he was crushed between two cars, resulting in death. See "Ernest Morris's Chum Wilkin Killed," *Indianapolis Journal,* December 10, 1880.

[11] Morris, *In Search of Human Heads*, Chapter I.

[12] "General News," *Titusville Herald*, June 26, 1870.

[13] "Personal," *Fayetteville Weekly Recorder*, December 29, 1870.

[14] "House and Lot for Sale in Manlius Village," *Fayetteville Weekly Recorder*, September 4, 1873.

[15] The Rio Maicá, in the flood plains east of Santatem, State of Pará, Brazil. Note that Morris's manuscript was created using a typewriter incapable of lower case letters or diacritics. In many cases, he underlined letters to be capitalized, and some accent symbols, but was not consistent. Believing that a copy editor of his time would have corrected these, those diacritics have been added.

[16] Preferred as *Atta cephalotes*

[17] A container for liquids varying from cup to pitcher-sized, made from gourds of the calabash tree.

[18] Often used specifically to refer to the starch derived from the cassava plant, *Manihot esculenta,* but commonly used in Brazil to refer to the plant itself. The derived starch is also known as tapioca. Morris uses mandioca to refer to the plant, and tapioca to refer to the starch. Another name for mandioca is manioc.

[19] Preferred as *Manihot esculenta*

[20] A tubular wicker press used to remove moisture during the preparation of farinha and other starchy extracts

[21] Ernest T. Morris, *In Search of Human Heads.* Internet Archive, accessed February 4, 2017. https://archive.org/details/InSearchOfHumanHeadsErnestT.Morris

[22] Known today as Lago Grande do Curuaí

[23] Morris, *In Search of Human Heads*, Chapter II.

[24] Preferred as *Victoria amazonica*

[25] "Sweets of Science: An Amazonian Conquest," *Indianapolis Sentinel*, April 11, 1876; "A Hoosier for Luck," *Atlanta Constitution*, April 18, 1876.

[26] Barrett Paine, the prospective companion for Ernest's canoe journey down the Mississippi, was involved with another famous, but controversial, canoe expedition. In July, 1881, Paine accompanied brothers Willard and Frank Glazier on an expedition intended to travel the full length of the Mississippi from its headwaters to New Orleans. At that time, the accepted source was Lake Itasca. However, from the 1830s through the 1870s, others had noted that a smaller lake, Elk Lake, fed into Lake Itasca, but did not try to publicize the fact. The Glaziers and Paine found the same lake (being led to it by Ojibwe guide), christened it Lake Glazier, and proclaimed that they had found the true source of the Mississippi. This prompted a vicious debate within the Minnesota Historical Society, with the Glaziers and Paine labelled as frauds and charlatans.

Barrett C. Paine had, by that time, fallen out with his father and others in the family, and after a few years in Minnesota moved to Brooklyn, New York, in 1890. For sixteen years, he edited a trade magazine for stationers, earning a reputation as a modest and respected member of that profession. Ironically, he died in 1906 at the early age of fifty-three from the effects of repeated bouts of malaria, dating back to his canoe journeys. See: "Obituary," American Stationer, May 16, 1906, p.41.

[27] "A Gifted Liar," *Indianapolis Journal*, April 14, 1876.

[28] "Morris Heard From," *Indianapolis News*, May 31, 1876; "City News," *Indianapolis News*, June 8, 1876.

[29] "Interesting Career of the Late John Akhurst," *Brooklyn Daily Eagle*, April 27, 1902, p.22.

[30] "At Home," *Indianapolis Sentinel*, June 8, 1876.

[31]Gilbert Motier Levette, the Assistant State Geologist of Indiana, who served as Ernest's mentor in the natural sciences, committed suicide in November, 1889. There were two theories concerning his suicide. The Indianapolis newspapers said that two years prior to his death, he had been in a bad streetcar accident and had fractured his skull, and had been moody and morose since that time, and unable to work. Those newspapers suggested he had taken his own life to avoid becoming a burden on others.[31] The Cincinnati newspapers suggested he was distraught over the fate of his son, whose behavior had landed him in the Indiana State Reform School. See: "Obituary," American Stationer, May 16, 1906, p.41; "Prof. Levette's Suicide, Indianapolis News, November 29, 1889

4 IN SEARCH OF HUMAN HEADS

Following the return from his first trip to South America, Ernest Morris remained in Indianapolis for less than two months. He then set out once again for New York City to catch a merchant ship bound for Pará, Brazil. Before leaving Indiana, the local newspapers briefly reported that he intended to spend about a year exploring more of the Amazon valley; and upon his return might next go to Africa.[1] He appears to have left on short notice, and gave no extended interviews detailing his plans. A former newsman for the *Indianapolis News*, Charles A. Tryer, was said to be accompanying Morris, but for unknown reasons did not leave town with Morris.[2]

Morris arrived in Brazil in late August 1876, having just turned twenty years old. On his first trip, he had success collecting specimens around the lakes, streams (*igarapés*), and channels (*paraná-mirims)* surrounding the western reaches of Lago Grande de Villa Franca; consequently, he decided to return to that area to begin his second trip. He took a steamer to the town of Óbidos on the Amazon, and from there a canoe to the village of Juruti, about 150 miles further up the Amazon from Santarem. Juruti was a village of about 150 people, and was immediately adopted by Morris as his favorite spot on the Amazon:

> It will be asked what are the advantages of Juruti? Simply its delicious climate which is probably one of the most perfect in the world. The air is always soft, cool, and balmy; there are no sudden changes; no damp, sultry days and hot nights. A cool breeze generally tempers the heat of the sun and the temperature is so unvarying that during a residence of many months we found the thermometer day or night to vary but little from seventy-seven degrees Fahrenheit in the shade. There is also a freedom from

insect pests, except during the months of August and September there are no mosquitos to annoy one. The rains are not violent or continuous and the nature of the soil is such that there is no standing water; and after the most violent showers during which the whole earth seems a river, one can in a few moments walk dry-shod in slippers. Notwithstanding this the country is never parched. The foliage is always fresh, bright green, and many very beautiful and brilliant flowers abound. There is no malaria, nor are there any prevalent diseases. Bright cloudless days are succeeded by clear starry or bright moonlit nights. One day is the counterpart of another and all are beautiful.[3]

Mouth of a Lateral River on the Madeira. by Franz Keller. From: Keller, Franz. *The Amazon and Madeira rivers; sketches and descriptions from the note-book of an explorer.* Philadelphia: J. B. Lippincott & Co., 1875. [Public Domain]

Morris spent the next three months collecting near the shores of the lakes in the vicinity of Juruti: Lago do Jara; Lago Curumucuri; Lago do Juruti Mirim; and Lago Grande de Juruti (where the old village, Juruti Velho, was located before the populace moved to the current Juruti). He never tired of finding new sensory delights in the region:

The fragrant air of these tropical woods is a constant source of enjoyment. The odors in one wood path will remind one of

> wild azalea; in another of the spicy Missouri currant; in another of Clethra; and again the breeze will seem to have blown over fields of white clover. At night also the whole air in the village would be full of delightful fragrance. Whence these odors come it is almost impossible to discover' probably, however, not from any large flowers but from the small flowered trees, great numbers of which are found in the Brazilian forests. Often when out a night, when not a breath of air was stirring, and nothing broke the silence except the cry of some lone night bird, a low murmur seemed to come from the forest—a monotone, soft, soothing, continuous, perhaps the ripple of multitudinous leaves united in one wave of sound...
>
> ...One morning when far beyond these plantations and an immense *roga* or clearing planted with mandioca, we saw a sight which will be long remembered. In a comparatively open space in the forest stood a large tree, the trunk a mass of short woody protuberances. It was loaded with small white flowers. Around this flew thousands of bright butterflies—red, blue, yellow, green, white and black. As they flashed in the sunlight the effect was most dazzling and reminded one of a giant kaleidoscope.[4]

As expressed in Morris's later written account of these months, he became restless and longed to explore a different landscape—someplace where other naturalists had not yet tread; where new varieties of flora and fauna might be found; and where the native peoples lived as they had before their culture had been disrupted by Western religion, industry, and customs. From his earlier journey of 1876, Morris recalled seeing distant blue hills rising south of Santarem, in the highlands east of the Tapajós River. He later wrote:

> The desire to explore them grew upon us day by day, but when we mentioned the wish to friends it was ridiculed as rash and foolhardy, if not impossible of accomplishment. We were repeatedly told, we should never return alive; that the interior campos were inaccessible; that savage Indians filled the woods and that disease in every form lurked in the forests. Yet opposition only fostered the desire and, rash as the undertaking was, we resolved to attempt it. Imagine a boy of nineteen[5], of by no means a vigorous constitution, wholly ignorant of the language and unacquainted with the customs of the people he must meet, possessing no suitable outfit for the journey, starting upon an expedition which at the shortest must consume many months, to

an unexplored region, and the attempt indeed seems rash.[6]

While Morris's account asserts that he was first motivated to journey up the Tapajós river purely by curiosity and the spirit of exploration, other evidence suggests that he had a specific mercenary goal in mind from the very beginning: to obtain a set of dried human heads taken as trophies by the Munduruku Indians living in the campos region between the Tapajós and Xingu rivers. There is one obvious reason why Morris dissembled in his retelling of events—he wanted to obscure the fact that he had foreknowledge that trade in these artifacts was illegal.

The government of Brazil had outlawed trade in human head artifacts decades prior to 1876. Not only did the federal government want to eradicate the traditional warrior culture of several indigenous tribes; there was also a concern that a demand for these objects by collectors both within and outside of Brazil would escalate the gruesome practice. Sale of human heads was therefore prohibited. When heads were occasionally confiscated, the official pronouncement was that the objects were decades old, pre-dating the ban on these artifacts; the government was reluctant to admit that the unauthorized practice still existed in isolated regions of the country.

In Ernest Morris's journeys on the Amazon through to the end of 1876, he had only experienced the indigenous population in the towns and villages that, in the chauvinistic term of that period, were considered "civilized." This was understood to mean that there was a Catholic church or mission in the vicinity; or a government or military official in residence; or that a large segment of the populace was engaged in the organized production of trade goods: rubber; mandioca; sarsaparilla; guarana; etc. The ethnic natives and *mestiços* living in these "civilized" areas were detribalized (although differentiated by some local customs), and by the 1800s spoke língua geral, a blend of Portuguese and the old Tupi language--even though that language had not originated in the Amazon interior.

The natives that Morris met in Santarem and Juruti were of mixed ethnic background, a consequence of centuries of Portuguese colonization and mass exterminations caused by European diseases. As Ernest Morris travelled up the Tapajós river via steamer from Santarem, he found the same situation in the villages of the lower section of the river, like Boim and Aveiro. These towns contained a larger mix of Mawé and Munduruku speakers, for he was entering the traditional homelands of the Sataré-Mawé and Munduruku tribes. He also passed many isolated thatch huts that belonged to native rubber collectors, since the Tapajós was famous for its high-quality latex; it was in the highlands above Boim on the Tapajós that the rubber-tree thief, Henry Wickham, had collected his 70,000 seeds for delivery to England.

Steamers could ascend the Tapajós no further than the town of Itaituba, which sits about 300 miles south of the mouth of the northward-flowing Tapajós. Beyond Itaituba, a series of more than a dozen cachoeiras, or raging cascades, prevented all vessels except canoes from traveling on the 250-mile length of the upper Tapajós. Ascending the Tapajós with a canoe past this series of rapids required painstaking portages; poling against strong currents; and dragging the canoe with ropes anchored to rocks and trees. Descending the rapids, though ten times quicker than ascending, was extremely dangerous. Many stretches of this section contained whirlpools, churning whitewater, and boulders jutting up in midstream. Each of these cachoeiras had been given its own name and character; an experienced Tapajós canoe pilot knew them as well as a Mississippi steamboat pilot knew that river's channels and sandbars--and such knowledge was highly valued.

Ernest Morris arrived in Itaituba in late November 1876, during the onset of the rainy season, which usually lasted until April or May. He was advised to not to attempt an ascent of the river for six more months, since the rising water levels made the struggle past the rapids even more difficult. Morris dreaded being stuck in Itaituba for that length of time, since it offered limited opportunities for specimen-collecting. In Itaituba, he had hoped to encounter more Indians following their traditional customs, but noted that the closest Sataré-Mawé village was a three-day journey into the forest southwest of Itaituba. Still, he met several members of that tribe in Itaituba, and noted:

> They are a civilized tribe, peaceable and industrious, their chief occupation being the cultivation and manufacture of guarana.[7] The men are of medium stature, with regular features and an intelligent expression; and the girls are the most beautiful we have seen among the Indians in all our journeyings.
>
> Their houses are long, high, the walls generally of *palma*, that is, of palm fronds, contain but a single room and are very dark inside. They are clean and neatly kept and are usually surrounded by fruit trees and clearings in which guarana is grown. The guarana (*Paullinia sorbilis*)[8] belongs to the natural family *Sapindaceae*. It is a climbing shrub with compound leaves and

axillary racemes of white flowers. The fruit is a pear-shaped, three-celled capsule with thin partitions opposite, which it splits open when ripe, each cell containing a solitary dark-colored seed about the size of a bean. The fruit, which is the part from which the guarana is manufactured, ripens in October and November. The seeds are gathered and carefully dried in the sun, in the manufacture of the best quality being also picked over and all unripe or imperfect ones taken out. They are then reduced to a fine powder in a hard wood mortar. Water is added till the mass attains the consistency of a stiff paste. This is made into cylindrical rolls about ten inches long which usually weigh about a pound, and is also molded into the form of birds, fishes, and into many curious shapes. These rolls are then dried in a *forno* and sometimes afterwards suspended in smoke and become excessively hard.

In the month of January, traders come to Itaituba to purchase guarana; many from Cuiabá, a city in the great province of Mato Grosso, above the falls of the Tapajós, where great quantities of it are consumed. The guarana made in the vicinity of Itaituba is considered superior to that made on the river Maues, as greater care is used in its manufacture. Guarana is extensively used in Brazil as a beverage and as a nervous stimulant; and on the Tapajós very generally takes the place of tea or coffee, and is considered especially efficacious as a preventive of disease. To prepare the drink it is grated, usually on the tongue if a pirarucu, to a very fine powder. A teaspoonful of this is then put into a small glass and sugar is added and stirred with the powder. The glass is then filled with water and then carefully stirred until the sugar is dissolved. A portion of the guarana is dissolved, but the dregs are never drunk.

The Indians on the Tapajós never travel without a supply of guarana, and drink great quantities of it simply mixed with water every day. Without sugar, it is bitter and not palatable, but sweetened it has a very agreeable taste and is a pleasant beverage. Its strengthening powers are wonderful, and one can travel for days with no other sustenance than guarana. Its active principle is a substance called guaranine, identical with the theine of tea, but no other known substance yields it so abundantly, the amount being 5.07 per cent as against good black tea which yields 2.13 and coffee from 0.8 to 1.00. The production of guarana varies

> greatly in different years and this controls the price; it usually sells in Itaituba at from forty to eighty mil-réis an arroba.

Fortunately for Morris, a prominent citizen of the lower Tapajós named Silverio de Albuquerque Aguiar Leverger sent an invitation to Morris to join him immediately at his island plantation located just below the first set of rapids. Leverger had also played host to naturalist Charles F. Hartt about six years earlier. Morris eagerly agreed to the offer. He later wrote about the excitement he felt on departing Itaituba aboard a 30-foot native canoe, accompanied by native paddlers:

> It was a glorious moonlit night when we left Itaituba. Perched on the *tolda* of the canoe we watched the Indian crew, all Mundurucus, ten in number, five for each side, all naked to the waist, their tattooed faces, their punctured ears, in which they wore blocks of wood, and painted bodies forming a striking picture in the moonlight. Sitting by our side was the owner of the canoe, a long, bark-wrapped cigarette stuck behind each ear, blowing a farewell blast upon a *buzina,* a sort of trumpet made of bamboo, which emitted loud shrill blasts which effectually disturbed the repose of the inhabitants of the town. No canoe on the Tapajós is complete without its *buzina*, with which its arrival and departure are always signaled and which serves to salute every house which is passed upon the bank. Some of these instruments can be heard for miles.
>
> How nicely the Indians poled! When the water was too deep, taking to the large round paddles, keeping most perfect stroke, now long, now short, quick or slow as set by the bowmen on either side, until at a shout from the steersman they would drop their paddles and instantly resume the poles, the motion of the boat not slackening in the least.[9]

Morris's approach towards the Indians of Brazil must have been influenced by recent events involving American Indians in the United States. In 1874, gold had been discovered in the Black Hills of Dakota Territory, causing a flood of miners and settlers to encroach on Sioux tribal lands. The United States government proved ineffective in maintaining its previous treaties with the Sioux to restrict these intrusions, and instead tried to make the Sioux agree to relocate to Oklahoma. The Sioux refused, and hostilities against them were commenced by the U. S. Army in early 1876. In late June 1876--the same period when Morris had briefly returned to Indianapolis after his first trip to Brazil--the Battle of the Little Bighorn occurred; the rout of Custer's poorly-led troops galvanized the white population against the indigenous peoples of the United States and its territories.

Because of this and previous skirmishes with other tribes, the negative portrayal of American Indians by white society that had been pervasive since the onset of European settlement turned even darker. They were frequently depicted as violent savages: ugly, unclean, murderous, drunken, and practitioners of barbaric customs. A prominent feature of this characterization was the notion that many tribes conducted ritual warfare and engaged in taking scalps as trophies. While a few native nations were engaged in constant hostilities against one another, scalp-taking was not widespread and was not limited to one ethnic group. In fact, as a practice it may have been promulgated more by European settlers than by American Indians. Following the Battle of the Little Bighorn, the star of dime novels and western-themed stage theatricals, Buffalo Bill, was recruited as an Army scout and vowed revenge. Two weeks later, he shot and then scalped a young Cheyenne warrior, boasting, "The first scalp for Custer!" Buffalo Bill later incorporated the event into a stage production and displayed the scalp as his own trophy.

Ingrained from a young age in this culture of fear, fascination, and misunderstanding of American Indians, Ernest Morris now encountered the Munduruku of Brazil. As he travelled up the Tapajós, Morris went from meeting detribalized natives that might claim Munduruku heritage to contact with full-blooded Munduruku who followed many of their original traditions. One defining characteristic of Munduruku culture was (and still is) prowess with handling a canoe—a skill that Morris understood and respected, having navigated the length of the Mississippi in his own canoe. When it came to maneuvering the large, heavy, twenty-to-thirty-foot canoes used in the Amazon valley, Morris was content to let native crews navigate and was in awe of their stamina and efficiency.

After a day and a half, the canoe carrying Morris reached the island residence of Senhor Leverger. Leverger was an elderly, wizened mulatto who maintained a plantation of between fifty and seventy-five Indians, most of whom he employed to collect rubber latex, paying them in food and other goods. At Leverger's Morris saw first-hand the labor conditions that underpinned rubber production, a yoke of inescapable servitude that latex collectors bore for the benefit of the rubber traders. The process of collecting rubber demanded hours that could no longer be spent in hunting game or in tending fields. Morris described the routine:

The season for working the rubber is on the Tapajós from May to December. This is the dry season. The sap also runs the other months of the year, but the rain mingles with the milk in the cups and ruins it. It is also well to allow the trees a season of rest lest they become exhausted. The first process is to scrape the trunks of the trees and remove all dirt. This is done by the women and children. A band made of pith of the Mariti palm (*Mauritia flexuosa*) is then coiled around the tree in the form of a trough, beginning at a height of six feet and continuing nearly to the ground. This trough is supported on long pegs driven into the tree and the lower edge is joined to the trunk of the tree with mud, which soon becomes very hard. On the ground at the foot of the trough is placed a small vessel holding about a pint, usually of clay, but sometimes small china vessels or tin cups made for the purpose are used.

The bark of the tree is then bruised in about a dozen places by a smart blow from a small iron instrument, but the bark of the tree is never broken through; and, statements to the contrary notwithstanding, cuts are never made into the wood. The milk, which is white, sweet to the taste and not unpalatable, collects in the cups and each morning is gathered by the women who go from tree to tree carrying a balde, made of the fruit of the calabash tree hollowed out, with only a small hole in the top by which it is filled and emptied and into which the fingers are inserted in carrying it. The yield of milk varies greatly, depending upon the age of the tree and the frequency with which it has been tapped. An old tree yielding but a few spoonfuls, while one tapped for the first time may give more than a pint each day.

The full baldes are then taken to a shed and the milk emptied into large pans. It froths and in appearance exactly resembles cow milk. In this shed a fire is burning made of the nuts of the Inaga palm (*Maximiliana regia*[10]). Over this fire is placed a funnel-shaped earthen pot with an orifice at the top about two inches in diameter, from which rises a dense white (not black as some travelers say) offensive smelling smoke. Small thin paddles resembling in shape those used in a canoe are held over the pans and the milk is poured upon them from a cuia. They are then held for a moment in the smoke being rapidly turned and a thin coat of rubber immediately hardens upon them. This operation is repeated many times until the weight of the rubber on the paddle is from five to eight pounds. The paddles are then placed against the side of the shed for a couple of days and a thin, watery liquid drains from them; after which the rubber is removed by cutting and is ready for market. In this state it is known as *borracha* or *seringa fina.*

India-Rubber Manufacture on the Banks of the Madeira. by Franz Keller. From: Keller, Franz. *The Amazon and Madeira rivers; sketches and descriptions from the note-book of an explorer.* Philadelphia: J. B. Lippincott & Co., 1875. [Public Domain]

The rubber industry broke down tribal and community bonds, and often caused a state of permanent indebtedness to the rubber-trader overseers. Senhor Leverger also owned black household slaves; and Morris noted Leverger's frequent physical abuse of some of them. Despite this backdrop of cruelty, Leverger treated Morris and other foreign visitors with gracious hospitability, and was a respected and valued friend to the canoe crews and traders that passed his island at the foot of the cachoeiras.

Morris stayed on Leverger's island for three weeks. On December 4, 1876, he wrote a letter to his father that was later reprinted in the *Indiana*

State Sentinel. The contents of the letter were boastful and out of character with Morris's later written account of his Tapajós journey; and directly contradicted all later accounts he gave concerning the timing of when he decided to seek out and trade for the trophy heads. In the letter written at Leverger's, Morris plainly states: "I go to collect the heads of the wild Parrentins [*sic* Parintintins]. These heads are severed from the body and preserved by the Munduracas [*sic* Munduruku]."[11]

This statement contrasts with a manuscript that Morris wrote several years later, in which he stated that he did not decide to trade for trophy heads until he was already far up the series of cachoeiras on the Tapajós, near the mouth of a tributary, the Rio Crepori, where he encountered a canoe full of Munduruku warriors out looking for their enemies, the Parintintins. In that same letter posted from Leverger's, Morris also wrote that his goal was to reach the Munduruku villages in the campos region east of the Tapajós, and would access that area by following the Das Tropas river, which flowed into the Tapajós. Clearly, his intention to visit those isolated[12] Munduruku villages was formulated before he started up the Tapajós rapids. His later revision of the timing of this decision was likely made for one reason: to disguise the fact that he already had been told that trade in these human heads was prohibited. In other words, Morris knew that what he planned to do was illegal before the fact, not afterwards. Senhor Leverger, for example, would have told him that.

Why Morris made collecting trophy heads the focus of his quest remains a mystery. It can be assumed that some unknown person (or persons) assured him that there would be a market for the head artifacts if he were to bring them back to the United States. On his first trip to the Amazon, Morris had collected beetle and butterfly specimens on behalf of two clients in Brooklyn: John Akhurst and Dr. Robert G. Eccles. Subsequent events suggest that it may have been Elias Lewis, Jr., the museum curator and director of the Long Island Historical Society in Brooklyn, who urged Morris to collect the heads from the Munduruku. Throughout his life, Morris evidenced a lack of any purely mercenary inclinations; but the desire to impress and earn the respect of a curator of a major natural history museum might have persuaded him to temporarily set aside beetles, birds, and butterflies and instead go after these grisly ritual objects.

Even though the waters of the Tapajós river were already rising due to seasonal rains, Morris met with luck in finding a trader named Cândido Pinto who planned to head up the rapids after stopping at Leverger's.

Pinto was returning to his home on the upper Tapajós, bringing with him a cargo of cachaça rum and *cuias*--gourd containers. Pinto, a dark-complexioned *mestiço,* employed a crew of eight detribalized natives, colloquially known as Tapuyas, manning a twenty-five-foot canoe. Senhor Leverger urged Morris to wait until late Spring, but Morris was anxious to leave. The dour Pinto was less than enthusiastic to take on a passenger, but relented.

More than a third of the manuscript that Morris later wrote about his first two Amazon journeys details the epic struggle involved in ascending the series of rapids of the Upper Tapajós to the mouth of the tributary Rio Cururu; and up the Cururu to the campos highlands area where the traditional Munduruku villages were sited. The first rapids they encountered were called Maranhãozinho--the little tangle--which was passed in seven hours by dragging the canoe upstream via ropes. To do this, Morris and the crew waded through neck-deep water, drawing the vessel behind them, and anchoring the rope on trees and rocks. The next day they tackled the more difficult Maranhão Grande, which necessitated a portage of a half-mile up a steep, rugged path. All cargo had to be carried on the backs of the crew, while the heavy canoe was pulled over land on log rollers, all accomplished in the drenching rain. That evening they camped at the foot of the next cachoeira, Coaita (sometimes spelled as *Quatá*)—the spider monkey. There they were beset upon by swarms of mosquitoes.

To get above the Coaita cascades, all the cargo was once again carried through thick forest, while the empty canoe was pulled by rope past whirlpools and churning water. The vessel was then reloaded with cargo, and pulled another 500 yards to the base of the next rapids, named Apuí after a genus of Ficus tree found on the river banks at that location. Getting past Coaita had consumed a whole day. For dinner, Morris and others volunteered to catch fish, an episode that marks one of the few instances in which he mentions one of the popular dreads of the Amazon, piranha:

> We also saw great quantities of the Javari palm (*Astrocaryum jauari*) which were full of its bright yellow hard fruit. Gathering some of these, and taking fishing lines, we went to the side of the pool. Beating the water with a pole to attract the fish, we used the palm fruit for bait; the piranha gathered around in great numbers and were easily caught, but not without the loss of many lines which were bitten through by their sharp teeth. These fish made

> an agreeable variety to our dinner, after which we took our cigarettes and a siesta on the sand *in puris naturalibus,* all our clothes, which had been thoroughly soaked, hanging on the bushes in the sun to dry.

In his several journeys to the Amazon, Morris bathed nearly every day; yet had no personal horror tales to tell of the razor-toothed fish. He had heard tales that others related about frenzied attacks by masses of piranha in response to blood in the water. Compared to insect bites and fevers, Morris considered piranha an inconsequential danger.

The entire next day was spent pulling the canoe past the Apuízinho rapids, the men jumping from rocks to riverbank and into waist-deep water. Morris ruined one of the two pairs of shoes he had, and decided thereafter to go barefoot, like the Indians. His feet soon were cut and bleeding. A similar day of work was needed to get past the next rapids, Uruá (shell). The men staved off hunger and exhaustion by drinking caffeine-rich guarana tea. The eastern shore of the Tapajós near Uruá and fifty miles above it was considered as territory inhabited by hostile Parintintin Indians, so Pinto's crew were obliged for the next few days to camp on islands in the middle of the two-mile wide river.

They avoided the next cachoeira, Tamandua—the anteater—by moving up the quieter water nearer to the eastern bank of the river. They did not want to get too close to the bank because of the possibility of attack, so instead of disembarking and using ropes, they poled the canoe forward. On occasion, they entered water too deep for their poles, and the current swept them backwards, so that they had to work to cover the same distance twice. At length their progress became easier, and they passed another minor set of rapids, Boburé, without any difficulties other than being bedeviled by clouds of *pium,* the biting blackflies that vexed Morris more than any other pest in the Amazon.

After spending a night in a shack of a lone rubber trader, Pinto and his crew paddled past the mouth of the river Jamanxim, finally beyond the danger of the Parintintins. They now employed a new method to progress against the current: using poles with hooks on their end, they grabbed onto forward trees and bushes and pulled the canoe forward. In this fashion, after two days they reached the landmark of Fecho dos Montanhas, or Close of the Mountains, where the river, which had been about a mile wide, suddenly contracted to a channel of only 200 yards. In this channel was the cachoeira of Acará, which was brief but troublesome to pass.

While struggling past these rapids, they were met by a group of Mawé Indians, armed with bows, but friendly enough in demeanor.

First Settlement of an India-Rubber Collector. by Franz Keller. From: Keller, Franz. *The Amazon and Madeira rivers; sketches and descriptions from the note-book of an explorer.* Philadelphia: J. B. Lippincott & Co., 1875. [Public Domain]

The next afternoon, Pinto's canoe reached the foot of the mile-long Mangabal rapids, where the men camped, hunted, and feasted on monkey meat. Mangabal was traversed by wading in neck-deep water and pulling the canoe forward; the men were encouraged by the fact that this was the last of the lower set of cachoeiras on the Tapajós. A day later, they paddled relatively calm waters a distance of forty miles, stopping just short of the only sizeable settlement on the upper Tapajós, the mission village of Bacabal.

Pinto desired to paddle past Bacabal without stopping, for reasons that Morris learned later. At first sight, Morris saw Bacabal as representing the last outpost of the civilized world, populated by 500 Munduruku, presided over by a fellow Christian--albeit a Catholic monk. Bacabal had been founded seven years earlier, in 1870, as a Brazilian government-sponsored mission for the uncivilized Munduruku. An Italian Capuchin friar named Pelino de Castrovalva was charged with establishing the mission. Originally, he was to have progressed further up the Tapajós, but stopped at the first Munduruku *maloca* (longhouse) that he encountered past the lower set of cascades. There he convinced the Indians to build a school and new houses for his mission.

Unknown to Morris, *Frei* Pelino was a figure of great controversy, a personification of moral ambiguity that would have fascinated a writer like Joseph Conrad. The Capuchin's first actions were to place himself as an intermediary between the Munduruku living in the region and the independent rubber traders and other river merchants (like Cândido Pinto). No trader could stop at Bacabal without Pelino's permission; no goods could be exchanged without his presence; and nothing taken out of the region's forests without his approval. To his detractors, which included many of the Munduruku living within his influence, Frei Pelino was a petty tyrant, who ruled by fear and controlled all the economic activity generated by the Indians: rubber collecting, sarsaparilla harvesting, rubber collecting, and extraction of copaiba[13] and patawa palm oil.

Morris was intrigued by the sarsaparilla production, and made notes on its processing and economic significance:

> The name is derived from the Spanish *zarza*, a bramble, and *parra*, a vine. It is the root of several species of *Smilax*, and on the Tapajós we noticed two species. They are tall, climbing, prickly vines, and are usually found in the lower lands and along the rivers and *igarapés*. In size the stem is about the diameter of one's finger, but one of the species we saw, and that of which the root was considered the better, was much smaller than the other. These plants climb very high in the trees and seldom branch or have foliage on the lower portion. In foliage, flower, and fruit the plants much resemble the *Smilax* of the United states.
>
> In gathering, the roots are dug, and, as they often penetrate to considerable depth, the labor in a close-matted forest, where the ground is a mass of roots, is very great. The roots are small and fibrous, but the root stock at first runs directly downward about a foot before the fibers spread from it. These root stocks with the fibers are the sarsaparilla of commerce. To prepare them, they are dried in the shade and then done up in long bundles, in which state they are sent to market.
>
> The larger species of the Tapajós sarsaparilla, which is probably *Smilax papyracea,* we have since seen on the Rio Negro, but have never met with the smaller species except on the Tapajós. Sarsaparilla is in Brazil called *salsa* and has much

> reputation as a remedial agent.[14] The root being mashed, soaked, and boiled for a long time until the infusion has the color and consistency of thin molasses. This plant occurs in great abundance on the Tapajós and some of its tributaries, especially the Rio das Tropas, Cururu, and São Manuel, but on account of the danger from hostile Indians and the extreme unhealthyness of the regions in which it grows, there is little exported, although the gathering would be exceedingly profitable.[15]

Despite his botanical curiosity, Morris could not overlook the tense social conditions of Bacabal. Much later, he learned from a knowledgeable source, Dr. Adriano Xavier de Oliveira Pimentel, that several Munduruku had made a long journey to the state capital of Pará to complain to federal officials about their virtual enslavement to Frei Pelino. They received vague promises that an investigation would be made; on facing the prospect of going back to Bacabal, two of the natives were so distraught that they threw themselves under the paddle of the returning steamer. Rumors were circulated that Pelino had taken a native mistress. Frei Pelino was recalled by the government to explain himself. In his own defense, he claimed that it was the region's rubber traders who had exploited the Indians; and that he had put a stop to it; and that the traders were now trying to defame him. He was permitted to return to Bacabal to continue his mission, but with closer supervision.

Ernest Morris was unaware of this state of affairs when he asked Pinto to stop at the village to call upon Frei Pelino. Upon arrival at the village, Pelino, wearing his gray cloak and sporting a long dark beard, greeted Morris as a friend, and had a generous feast prepared. Later, Frei Pelino gave Morris a tour of the mission; he seemed especially proud of the large storerooms of salsa, rubber, and containers of oil. However, Morris also spotted several men chained and locked in stocks—they were river merchants who had run out of food and made the mistake of trying to negotiate a deal with villagers without first consulting Pelino.

Morris was not so flattered by the good manners of Frei Pelino that he failed to notice the poor state of Bacabal's Munduruku:

> Missionary work on the Tapajós, as shown by the settlement at Bacabal, has done little real good to the Indians. They learn all the vices of the white man without acquiring his virtues, become idle, thieving, drunken and shiftless; and the Mundurucus living at the mission at Bacabal are in every respect inferior to their so-called "less civilized" brethren who inhabit the malocas further up the river.

With some relief, Morris, Cândido Pinto, and the crew of paddlers left Bacabal the next morning and proceeded upriver fifteen miles to a point where two great hills loomed over the western bank of the river: Quataquara (house of monkeys) and Santa Barbara. There Morris spotted ancient hieroglyphics etched into the cliff rocks—the first evidence that he had seen to substantiate a fable he had been told of back in Brooklyn: that a great ancient civilization had once existed in the Amazon. Morris's interest in the carvings was only mild; archaeology was not one of his passions.

The next morning, they passed the mouth of another tributary, Rio Crepori. As they moved by, another canoe shot out from the smaller river—a war party of Munduruku from the interior. They were on their way down to the river Jamanxim to search for their enemies, the Parintintins. The war party canoe maneuvered over to Pinto's canoe, where they exchanged news with Pinto's Munduruku crewmen. In Morris's later written account, it was during this exchange that he first heard that trophy heads were still taken during these skirmishes, and Morris decided at that moment to go and obtain some of these artifacts. This contradicts statements he made weeks earlier in a letter written at Senhor Leverger's.

Shortly afterwards, Pinto stopped the canoe to allow his crew three day's rest: many among them, including Morris, were suffering from chills and fever, likely victims of an outbreak of malaria. After that much-needed respite, just a day later they arrived at Cândido Pinto's *barraca* home on an island in the Tapajós, a half-mile from the mouth of another tributary, the Das Tropas. Morris had no plans on how he was to proceed from Pinto's island, other than vague assurances from Pinto that Munduruku from further up the Tapajós or the Das Tropas sometimes stopped at his *barraca* to trade. Despite the pleasant surprise that Pinto's usual sour mood had turned gracious once he reached his home island, Morris found little to do there. There was hardly any flora or fauna to observe and collect, though he was delighted to find poerie, or singing fish, in pools of nearby water.[16]

Had Morris been patient, Munduruku from villages on the Das Tropas might have soon stopped at the island and offered to take him up that river, from which the villages of the highland Munduruku—the trophy-head warriors--could have been reached with relative ease. Instead, when Cândido Pinto mentioned that he was going to send a small canoe with goods to a Munduruku village further up the Tapajós, Morris asked to go with them. There were a few traditional Munduruku *malocas* left on the Tapajós, and all of them above Pinto's island. Though these Indians kept many traditional practices, they did not collect heads like their more isolated highland cousins. Still, Morris must have weighed the chance that these villages might have ways to connect him with the inhabitants of the campos, and so decided to tarry no longer at Pinto's barraca.

House of a Wealthy India-Rubber Collector by Franz Keller. From: Keller, Franz. *The Amazon and Madeira rivers; sketches and descriptions from the note-book of an explorer.* Philadelphia: J. B. Lippincott & Co., 1875. [Public Domain]

Morris and the few men that Pinto assigned to the canoe proceeded up the Tapajós for three days, stopping briefly at a several Munduruku villages. Though he wasn't a copious note-taker, he later described them succinctly:

> The Mundurucu Indians all live in malocas, so called, which are long houses about twenty-five feet wide, one story, with high-pitched roof, thatched with palm, the sides usually mud, and containing numerous doors and windows. There is but one maloca in each settlement. It is often 150 feet long and contains but a single room. In this all the people live; it is full of dirty hammocks swung in every direction, and often contains *fornos* for

> cooking. Outside of the maloca there are often long, open, palm-roofed sheds under which the Indians work by day and hold their dances by night; and a prominent feature of which is always an immense *gambá* or drum and huge bamboo trumpets.

Morris admired the survival skills and honest nature of the Munduruku, but could not overcome his repugnance to the smell and taste of their prepared foods; their body modifications (ear and lip blocks, body and facial tattoos); and the fetid smell of the oils and pigments they used for adornment and to ward off insects. However, he did his best to disguise his visceral reactions, and did not insult his hosts.

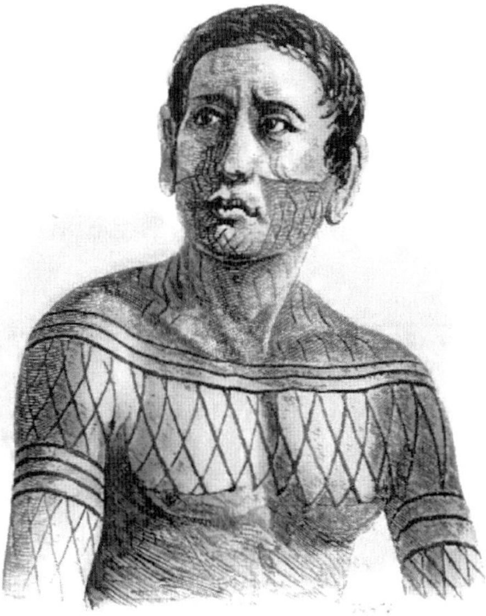

Indien Mundurucu by Édouard Riou. From: Biard, François. *Deux anness au Brésil.* Paris: Librairie de L. Hachette, 1862. [Public Domain]

After a few stops, they arrived at the foot of the first of the upper rapids of the Tapajós, the cachoeira of Chacarão. No paths existed to portage around this mile-and-a-half cascade, so once again an entire day was spent wading in swift-running water to pull the canoe forward. That evening Morris swung his hammock as a guest in the maloca of Chacarão; and the next day a few hours of travel brought them to their destination, the maloca of Antonique. The journey from Cândido Pinto's island had taken ten days. Antonique was the name of the *tuxaua*, or leader, of this maloca. Although the Munduruku had no single tribal chieftain, Antonique was said to have great influence among the other malocas on the Tapajós and with the sub-group of Munduruku in the highlands, the *Campineiros.*

Antonique was a tall, muscular man; and Morris, though usually unimpressed by body modifications, described his tattooing as beautiful. Antonique spoke língua geral and even a bit of Portuguese, and often journeyed down the Tapajós, and even beyond that to Pará. After Morris made his intention to visit the malocas in the campos known, Antonique welcomed him as a guest, provided a space in the maloca for him, and advised him to wait patiently until a canoe might be arrive or be readied for the journey up one of the eastern tributaries to the highlands. To pass time, Morris collected butterflies and observed daily life, but soon grew bored. He saw the men of the maloca frequently leave for day-long hunting expeditions, and asked Antonique if he could join them. Antonique refused, thinking Morris would slow them down and scare away game. Morris persisted, and finally Antonique allowed him to accompany a hunting party of six men.

Morris was armed with a custom-made shotgun crafted for him by an Indianapolis gunsmith. The others in the hunting party had bows and cheap-quality, poorly-maintained firearms. Before leaving the maloca, Morris followed the lead of the others and stripped to his waist; and then smeared his body and face with red *urucu,* a grease-like pigment, and black *genipapo.* The party paddled a canoe up a nearby small stream, and before heading into the forest, removed the rest of their clothes. They spilt into two groups, each moving in single-file through the hilly woods.

The image of twenty-year-old Ernest Morris, with his naked, painted body, moving silently through the rainforest and united in purpose with the Munduruku hunters, is one of the most memorable moments of his career. He later wrote:

> Suddenly we heard the report of a gun from the other party and in a few moments a large troop of immense coaita monkeys (*Ateles paniscus)* came towards us, swinging hand over hand from limb to limb in the tall trees. We crouched behind trees, fired, and three were brought down by the party, one a female with a young one on her back. This is the largest monkey found in Brazil, often when fully grown standing four feet or more in height. It has long, coarse, shaggy black hair over the whole body, but the face is bare and flesh-colored. The arms, legs, and tail are very long and attenuated. The flesh of this monkey is considered better eating than that of any other species, not only by the

> Indians, but is also esteemed a delicacy by the Brazilians. On the opposite bank of the Tapajós, another species is found, probably *Ateles marginatus*, resembling the former in every aspect except that the whole face is covered with short white hair.
>
> All day we wandered in the forest finding plenty of game, chiefly birds as: *araras* (macaws); white-throated toucans, a very large species; mutum, a large bird with glossy black plumage, large orange bill, and a crest on its head which it can raise or depress at pleasure; and other game birds. We also shot many cutias, a small rodent animal about as large as a rabbit, which has a white tender flesh; and another species of monkey called by the Indians cuchin, of a dark steel gray color, it has a bushy tail, the hair on the head is parted exactly in the middle, and it has long side whiskers which give it a very dandified appearance. The teeth are used by the Indians for necklaces, and the tail often serves as a brush. We reached the maloca at night having enough game to supply a large boarding house for weeks (provided the boarders were fond of monkey), but all was immediately cooked and devoured and the next morning nothing was to be had for breakfast except castanha mingau.

Morris stayed at Antonique's maloca for over a month, and though he made many observations, he wearied of the daily routine. Often, the mood was somber—a contagious fever had swept from village to village, and many people died at Antonique's maloca. Some suspected that Morris was the cause of the bad fortune, arguing that Morris was a witch, but Antonique defended the young American. Morris sensed he might be in some danger, but likely did not know how serious the charge against him could have been. To be identified as a witch often meant death by execution—perhaps the only instance in which Munduruku would harm one of their own.

At length, the trader Cândido Pinto sent a few men to Antonique's, with instructions to Antonique to send a trading party into the highlands to obtain artifacts—including dried human heads—from the Campineiros. It is not clear whether Pinto's intention was to assist or compete with Morris's trade intentions, but Morris was in no position to pass up the opportunity. Antonique, who also needed to collect a load of rubber latex from a maloca in the campos, prepared two canoes to venture up the Tapajós and then proceed up the Rio Cururu; but he did not want Morris to accompany them, instead offering to conduct the trade and bring heads back to him. However, Morris insisted on visiting the campos in-person—

and when Antonique tried to discourage him, Morris went and sat in one of the canoes until he relented.

The better part of three days was needed to get the two canoes past the cachoeira of Capoeiras, a three-mile long set of rapids. The assigned paddlers included women and children, far less skilled than the crews that Morris had traveled with previously. During the transit of these waters, Antonique and Morris were in the larger vessel; they lost sight of the smaller canoe, but later discovered that it had broken loose while being pulled forward and had been dashed against the rocks, killing a woman and two children. Also lost was the whole cargo of trade goods that Cândido Pinto had sent to exchange for artifacts of the Campineiros.

Another two days of struggle brought the remaining canoe to the maloca of Ireré, just above the head of rapids of the same name. There they met a stranded Brazilian rubber trader, who went by the name Barnabé. Barnabé had recently gone up the Rio Cururu to trade goods with the Campineiros in exchange for rubber, on the promise that they would bring it to him in Ireré. The delivery of the latex had not yet occurred, so Barnabé felt compelled to return to the campos to collect what was owed to him. Reluctantly, Antonique agreed that Barnabé could accompany them.

More advancement was made up the river by paddling and using hooks, with Morris joining in the effort since the crew was short-handed. Two long days later they reached another settlement, the maloca of Gabriel, the last of the Munduruku longhouses on the upper stretch of the Tapajós. The rains were violent and continuous, and Morris and some of the Indian crew still suffered from chills and fever. There they reinforced their food stocks, added a couple of crewmen, and then proceeded up the river, pursued by clouds of pium and mosquitoes.

A day and a half paddling brought them at last to the mouth of the river Cururu, now swollen from the seasonal rains. The Cururu was a narrow river, very meandering, with countless oxbows that doubled the river back onto itself. Its source lay in the highlands, where waterfalls prevented canoes from venturing—a further source of insulation for the Campineiros from the creeping effects of "civilization." Antonique's plan was to advance up the river for a few days to attain a landing known to both Antonique and Barnabé; and from that landing to follow a path to one of malocas of the Campineiros. Morris was stunned by the beauty of the Rio Cururu:

> The forests along its banks are magnificent, being far more lofty and exceeding in variety of trees any we had before seen. Many of the trees as we passed were loaded with orchids which were full of flowers, and the air was heavy with their rich, spicy fragrance. The variety of flowers, in which the colors white and pink predominated, was very great… We noticed many peculiar trees in the forest, indeed, much of the vegetation was wholly new to us, and we are utterly at a loss to describe it botanically.

As they moved up the Cururu, Antonique once again suggested that Morris should hand over his supply of trade trinkets, and that Antonique would go to a maloca and bring the dried heads back to Morris. Morris did not entirely trust Antonique, and declined that offer. It could be that Antonique judged that the Campineiros would not willingly trade any of the heads to a white man, and was trying to do Morris a service. Whether it was a misunderstanding or not, relations between Antonique and Morris grew worse. After three days on the Rio Cururu, one morning Antonique abruptly broke camp at the riverside and told Morris he was taking the men overland to the maloca of Arrakeray, a march of two days. He left Morris, Barnabé, and a teen Munduruku (who was too ill to walk) with the canoe to fend for themselves, with just a small supply of food. Antonique likely assumed that the stranded Morris would have no choice but to remain camped at that spot until Antonique returned or sent aid from Arrakeray.

Once the three were alone, Barnabé said that he thought they should turn the canoe around and return down the Rio Cururu to one of the malocas back on the Tapajós. The Munduruku teen had no strength to care one way or the other. But Ernest Morris, despite still having a fever himself, was maddened and even more determined to press onward in the face of Antonique's abandonment. He convinced Barnabé that they should continue up the river and search for the landing spot where Barnabé had many weeks earlier been able to follow a trail to a maloca of the Campineiros:

> It was a dark, gloomy morning, the sky thickly overcast with clouds, threatening instant rain as we embarked and paddled slowly up the river. The prospect was anything but cheerful and our only comfort was that the rain held off. All day the canoe crept up against the current. We often stopped to try to find high land where we might be able to shoot something for dinner, but both banks were flooded forest, thickly matted with vines, which seemed to extend for miles on either side, and we could not

> penetrate into them. Even where small openings appeared to communicate with what seemed to be tracts of open water—either lakes or flooded campo—our canoe was too large to enter. At night we tied the canoe to a tree, made a supper of farinha and water, and slept under the *tolda* of the canoe surrounded by swarms of mosquitoes.

For three days, the trio paddled slowly up the flooded Cururu, sometimes losing all sense of the direction of the river's flow, for they were surrounded by a flooded forest. The young American must have wondered if he reached the most isolated, alien place on the planet, a place where no one would ever look for or find him. Their attempts to hunt game met with failure, and their survival seemed perilous:

> We had early in the morning crossed the river towards the southwesterly shore where the land appeared to be higher, in the hope of being able to shoot some birds or game, but were wholly unsuccessful. Towards afternoon we came to a narrow strip of land which rose about a foot above the level of the water. Here we landed to walk about and stretch our limbs, which were cramped from long confinement in the canoe. While here we heard close at hand a shrill whistle which we knew to be an *anta* (*Tapirus terrestris*), for so the tapir is called in Brazil. Seizing our gun, we ran towards the spot and got a good shot at the animal. Although badly hit the wound was not immediately fatal and the *anta* rushed into the woods and dove into the *igapó* where it was impossible for us to follow him. After hunting for some two hours in the hope of finding either birds or monkeys, wholly unsuccessful we re-embarked and again crossed the river to the northwesterly bank.[17]

Head of a Swimming Tapir, Pursued by Dogs by Franz Keller. From: Keller, Franz. *The Amazon and Madeira rivers; sketches and descriptions from the note-book of an explorer.* Philadelphia: J. B. Lippincott & Co., 1875. [Public Domain]

Ultimately, with relief Barnabé signaled that they had arrived at their destination:

> The forest became more open and at last the canoe grounded. "Here is the port," said Barnabé. "Where?" we asked, for there was nothing but woods and water all around us, and as far as we could see through the trees, nothing but coarse grass growing in the water, and no sign of any opening leading to dry land. On further inquiry, we found that the canoe could go no further, and that to reach dry land and the path which led across the campo it was necessary to wade a long distance through the water.[18]

The Munduruku teen was still too ill to move, so Morris decided that the boy would stay with the canoe while he and Barnabé would try to reach the maloca and send help back. The water was knee-to-hip deep, so everything had to be carried on their backs using palm baskets they hastily constructed. They slogged through the water, occasionally getting their feet dry on small hills, all the morning and into the afternoon. Their stamina was quickly wasting to nothing. Abruptly, Morris was confronted with a danger more threatening than hunger and exhaustion:

> At last we saw before us a ridge of higher land which we found marked the course of the stream on its opposite bank. Towards this we directed our march, but were suddenly stopped by the deeper waters of the stream. Hearing a slight noise, we sank down behind a tree looking toward the opposite bank of the stream whence it proceeded, and beheld a sight which made our hair stand on end: six large naked Indians, their bodies painted red as blood, their heads shorn, and heads and faces painted in broad red and black lines, cords wound round their ears from which hung long feathers reaching to their waists, all armed with long bows and arrows.

Morris prayed that they were Campineiros. Crouching behind a tree, he called out in one of the few words of Munduruku he had learned, "Friend! Friend!"; and only showed himself, slowly, when the Indians responded, "Friend!"

---

[1] "Going Again," *Indianapolis Sentinel,* July 15, 1876.
[2] "Personal," *Indianapolis People,* July 15, 1876.
[3] Morris, *In Search of Human Heads,* Chapter III.
[4] Ibid.
[5] Morris was twenty at the time of his Tapajós journey, not nineteen.
[6] Ibid.
[7] Guarana has been one of the main ingredients of mass-marketed energy drinks since the late 1990s.
[8] Preferred as *Paullinia cupana*
[9] Morris, *In Search of Human Heads,* Chapter V.
[10] Popularly known as the Jagua palm, now preferred as *Attalea maripa*
[11] "Our Young Naturalist," *Indiana State Sentinel,* February 7, 1877, p. 3.
[12] The Munduruku of the campos villages were geographically isolated, but they did not shrink from contact. Indeed, their warrior culture was engaged with other peoples, as enemies, allies, and trade partners.
[13] Copaiba is an oleoresin derived from *Copaifera guianensis* trees. It is used as a traditional medicine.
[14] More specifically, sarsaparilla extract was used by peoples of the Amazon to treat impotence, weakness, rheumatism, and skin ailments, including leprosy.
[15] Morris, *In Search of Human Heads,* Chapter VIII.
[16] The colloquial name for this fish that Morris uses is not corroborated, but is likely the Silver Croaker, *Plagioscion squamosissimus*
[17] Morris, *In Search of Human Heads,* Chapter IX.
[18] Ibid.

5 THE TROPHY RITUAL

In 1877, American newspapers described Ernest T. Morris as the "first white man" to visit the Campineiros sub-group of the Munduruku--a claim that is troublesome both in its veracity and in its implied racism and chauvinism.[1] What may be more accurate to assert is that he was the first foreign-born naturalist to journey to their remote highland villages. By the 1870s, the Campineiros occasionally exchanged goods with outside traders, offering rubber latex (for which they had little use) and featherwork in exchange for practical tools, like knives and fishhooks. When Ernest T. Morris arrived at one of their malocas in obvious ill-health, carrying little more than one or two satchels, and accompanied by a rubber trader they knew, the Campineiros likely judged Morris to be an odd, but non-threatening *blanco* (white man).

Morris's visit came at a unique time in the history of the Munduruku people. Ethnologists currently theorize that a few hundred years ago, the original Munduruku moved into the territory east of the Tapajós, between the Rio Cururu and Das Tropas rivers, from the campos lands to the south, in the current state of Mato Grasso. Morris wrote that Campineiro chiefs told him the same migration tale.[2] Very late in the eighteenth century, after a few deadly skirmishes with Brazilian government forces, the Munduruku agreed to cease hostilities against non-Indians. They remained geographically sheltered from outside influences until the mid-1800s, when the rubber industry crept up the Tapajós, bringing individual traders and their western goods.

Some Munduruku abandoned tribal life and their homelands completely, moving down to the lower Tapajós and mixing with other de-tribalized Indians and non-Indians. Among those who remained, the villages along the upper Tapajós riverbanks adopted many of these outside influences, including: wearing cotton clothing; hunting with guns; drinking alcohol; eschewing tattoos; sometimes living in individual barracas rather than communal malocas; harvesting salsa, copaiba oil, rubber, etc., for trade; and, near the Bacabal mission, converting to Christianity. In Morris's day, these river-dwelling Munduruku were called "semi-civilized"; but Morris believed them to have been corrupted from their traditional culture, and in sad decline.

However, about a quarter of the Munduruku population inhabited villages in the campos highlands, located a score or more miles from the upper reaches of the Rio Cururu and Das Tropas rivers. These people, the Campineiros, were still conducting tribal warfare against neighboring tribes late into the nineteenth century. It was a practice imbued with ritual and symbolism, central to their way of life. They took the heads of their enemies as trophies, and dried and decorated them for use in dances and feasts. They were the most feared warriors in the Amazon basin. Often, they were the aggressors, and raided the territory and villages of their enemies. Their weapons were bow and spear, and sometimes cheap, poor-quality firearms. In some instances during the nineteenth century, the Munduruku were employed by Brazilian authorities to subdue other tribes that were threatening settlers and rubber traders; because of this, the Munduruku managed to continue their warrior culture into the early years of the twentieth century.

Around 1910-1915, their warfare ceased; they no longer had any competing tribes to battle, and the government could no longer tolerate their murderous raids. It was a loss of one of the central aspects of their culture, and created a void that for decades appeared to be pointing to their disintegration as a tribe. In recent decades, they have united with a new purpose—to defend their territorial homeland from encroaching development pressure applied by dam builders, gold and diamond miners, loggers, and ranchers. It was during a low point in their culture--the early 1950s--that the first detailed anthropological study was made of the Munduruku by Americans Robert F. Murphy and Yolanda Murphy. They had no knowledge of Ernest Morris's earlier visit in 1877, and no access to his brief notes. What the Murphys missed were Morris's documentation of some of the Munduruku practices surrounding war parties and the ceremonies in which the dried heads were employed.

The hunting party that Morris and Barnabé encountered brought the two back to their village, which Morris understood to be named Apopé, while other men were dispatched to aid the teen boy they had left back with the canoe, too ill to move. Morris collapsed in a hammock and rested the remainder of the day; in the evening, he was greeted by the chieftain of Apopé. Barnabé served as interpreter, and eventually Morris was given permission to trade with people in the village. As a new guest, Morris was taken to each of the four round houses in the settlement and given food at each as part of a welcoming ceremony. As hungry as he was, he had a visceral repulsion to the texture, smell and taste of the Mudurucu meals, and feigned a toothache so as not to insult his hosts.

Morris's stay at Apopé lasted just a few days, but was long enough to witness a feast and accomplish part of his goal: trading knives and fishhooks for five dried heads. He then accompanied a party who took him twenty-five miles through the campo, to the village of Carocopé. There he witnessed a daily mourning ceremony that no one else has documented:

> A few days previous to our arrival at Carocopé, a war party had returned, having lost some of their number in an encounter with their enemies; and the heads of the slain had been cut off and carried away. They had brought to the maloca the bones of the bodies divested of flesh, and that night a solemn mourning ceremonial was held over them. Just after dark a large fire was lighted in front of the *quartel,* and, summoned by the blowing of *buzinas*, the warriors, all painted red and black and wearing more or less feather work, gathered around. One of the chiefs was very beautifully dressed. On his head was a tall crown of the long and yellow feathers of the *arara.* Necklaces of teeth hung round his neck; from his ears hung long cords reaching to the waist, covered with bright feathers. A broad sash or rich feather work passed from his right shoulder to his left side, and descended to the knee. Around his arms and legs were many amulets and anklets of feathers, and around his waist was a broad girdle made of blue, white, yellow, red, and other colored feathers from the breasts of birds. In one hand he carried a large close plume of rich feathers and in the other a long spear. It is very difficult to procure a complete suit of this feather work, such as we have described, as few of the Indians wear them, and only at grand ceremonials. We, however, purchased this suit for one shirt, two razors, two knives, a looking glass, and twenty-five fish hooks.
>
> Arranged in a circle in front of the *quartel* stuck into the ground were all the weapons of their enemies which they had taken in battle: bows, arrows, spears and broken lances, some beautifully made, and many adorned with human hair. In the center of this circle, five human heads were placed on poles, their long hair waving in the night breeze and their dark features and waxen eyes showing ghastly in the flickering fire light. To the low,

solemn music of *buzinas* and the roll of the great *gambá*, the warriors marched slowly up and down droning a low chant, which seemed to consist of the two words "mama-mama," indefinitely repeated. Ever and anon as the drum sounded louder, one of the warriors would break from the ranks, rush towards the human heads, and go through war-like motions, brandish his spear in the faces and utter inarticulate exclamations.

Mundrucu Indian Woman. From Brown, Robert. *The peoples of the world: being a popular description of the characteristics, condition, and customs of the human family.* London: Cassell, 1882. [Public Domain]

Groups of women stood around, mostly naked, but a few who were the happy possessors of an article of raiment, wore it in honor of the occasion. It consisted only of a single piece of cloth or a short skirt which did not reach to the knees. The dancing and music continued until late at night, and when finally we went to sleep it was only to be roused by the loud roaring of a jaguar, which was prowling around the maloca. The Indians seized their bows and arrows and went out, but were not successful in killing him.[3]

At Carocopé, Morris obtained six more heads, including some of recent vintage. He was shown the smoking process, which depended on a creosote-rich root of a plant that Morris was told was the *carocopowpow*[4]. This root, more than the heads themselves, was difficult for Morris to obtain from the Campineiros. Ultimately, he obtained a large piece only by offering his own last, personal knife.

After a few days, Morris returned to Apopé, where he found that Antonique and his men had arrived. Morris still believed that Antonique had left him to die on the Rio Cururu; but, for his part, Antonique appeared to be genuinely glad to see that Morris was alive and had obtained the heads he sought. Since Morris knew he would have to depend on Antonique to return safely down the Rio Cururu to the Tapajós, he did not make an issue of Antonique's earlier desertion. Antonique informed Morris that they needed to return to his maloca.

The Campineiros took the news that Morris was leaving with obvious regret; and Morris himself expressed a desire to stay longer to collect specimens, since so much of the flora and fauna of the campos was unknown to him. The villagers marched with him and helped carry the dried heads along the long trail back to the landing on the Rio Cururu, swinging the artifacts from the ends of poles on their shoulders. At the canoe, the Campineiros called goodbye to Morris and asked him to return with knives and cuias to trade. Morris's fever had returned, and he rested under the canoe's *tolda* while Antonique and his men paddled down the flooded Rio Cururu. The air was fragrant with blooming orchids, more so than Morris had seen just a few weeks earlier during the ascent of the river.

On their second day of travel, the crew reached the confluence of the Rio Cururu with the Tapajós, which had now crested at its seasonal high water stage. Where they expected to encounter the first set of rapids, the cachoeira of Piscorina, they found the rocks were all obscured by the floodwaters. Since minimal paddling effort was needed due to the current, they continued through the night. Early the next morning they stopped at the first settlement of the river-dwelling Munduruku, the maloca of Gabriel, long enough for a good breakfast. By evening they had advanced down the Tapajós to the next maloca, Ireré, returning Barnabé to the place where Morris had first met him. Morris bade farewell to Barnabé, who had been indispensable to him.

From Ireré, Morris's return trip down the river turned into a thrilling paddling adventure story:

> The next morning, we embarked, but with a large addition to our passenger list, for there were many of the maloca who were waiting for an opportunity to go down the river, and the canoe was crowded from stem to stern with women, children, babies, and dogs, and thus rendered most uncomfortable. A short time after leaving the maloca of Ireré we arrived at the head of the cachoeira of Capoeiras. Here Antonique resigned the post of steersman to one of the Munduruku crew, who steered the canoe boldly out into the middle of the river. The six men on each side grasped their paddles, and one man armed with a long paddle took his station at the prow.
>
> All in the canoe were made to sit upon the bottom of the boat, in order that there might be nothing to obstruct the view of the steersman ahead, and also the better to trim the canoe. At a signal from the steersman the crew dipped their paddles into the water and impelled the boat forward with all their force, and in a moment we were in the midst of the rapid. The water foamed and boiled on every side. The roar was deafening and the canoe tossed up and down frightfully. The distant banks on either side seemed flying away from us and it seemed every moment as if we should be engulfed. Not a word was spoken, and the only sound, except for the noise of the waters, was the loud call of the steersman shouting his orders to the crew or to the man in the prow.
>
> In some places we saw great black rocks against which the water dashed furiously. In others, great *remoinhos*[5] showed the presence of huge masses beneath the surface, and both it was necessary to avoid, as to have encountered either would have been instant destruction. Fearful as the passage was, we could not but admire the presence of mind of the steersman and the skill with which he managed his immense paddle, and on him our safety depended. His features were perfectly immovable, his eye always fixed ahead. Often he threw the whole weight of his body upon the paddle, then again instantly recovering himself he stood erect like a statue holding it immovable by his powerful arms.

> Near the foot of the cachoeira an immense rock arises on either side. Between these the water rushes with immense velocity in a smooth unbroken sheet. To the east and west of these rocks, between them and the shore, the river is one mass of seething white caps. Towards this channel, which was only about twelve feet wide, the steersman directed the canoe. It was at once seized by the violent current and we instantly found ourselves with the rocks on either hand. A moment more, and we found ourselves in comparatively still water. The crew all threw up their paddles and raised a great shout. Never for a second had their paddling relaxed, and the sweat was pouring in streams from their bodies. In less than half an hour we had passed the cachoeira of Capoeiras, which it had taken us more than three days to ascend.[6]

An hour later, they reached the maloca of Antonique, where many residents were still suffering from fever, as was Morris. Morris hung his hammock in the communal house in the same spot he had many weeks earlier, too weak to think about how he would continue downriver. Antonique found his wife very sick, and she soon succumbed to the illness. To add to the misery, the rains continued incessantly. One day Morris checked his prizes--the dried heads--and discovered that some of the recently decapitated ones now had tiny worms. To save them, Morris had to take his *carocopowpow* root and use it as the Campineiros had, smoking the affected heads to complete the drying process. In a sense, the heads had now become Morris's own spirit trophies, imbued with his personal struggles no less than the warriors who cut them from their owner's bodies.

Morris recuperated at Antonique's maloca, while the grieving Antonique busied himself stockpiling a load of latex to transport downriver to Candidó Pinto's barraca. In early April 1877, he readied a canoe and presented Morris with the opportunity he sought. Morris later described the next part of the descent:

> We were soon again on our way, and at the head of the cachoeira of Chacarão. The upper part of this rapid is not dangerous when the river is full, but the incline and rush of water seemed to us very great, and we shot along with lightning swiftness. As we reached the lower part the water was broken into great sheets of foam and the canoe tossed up and down fearfully. There were but few rocks visible, but the danger was greater on this account, for now and then we could see the dark points as the water dashed against them and receded. We shipped much water,

> as the canoe was heavily loaded with rubber, *salsa,* and oil, but the Indians did not appear to regard the passage with any fear and, contrary to their usual custom when shooting a rapid, they kept up a continual laughing and shouting. We soon shot into the smoother water below and were obliged to land and take out some of the cargo to bail the canoe.[7]

Just one additional day and night of paddling through torrential rain brought them to Pinto's island, where Morris was delighted to find real coffee to drink and a place to sleep in relative quiet. However, when he awoke he found Pinto, Antonique, and all their men working furiously to build a floodwall around the house as the waters continued to rise. When those efforts failed, the most valuable items in the house and storage sheds were carried to higher ground. Morris lost all his flora and fauna specimen collections—all the seeds, leaves, soils, feathers, insects, etc. he had gathered since leaving Leverger's plantation and ascending the Tapajós rapids. However, his Campineiros artifacts, the heads and featherwork, were saved.

When the weather cleared, two large canoes were loaded with all the rubber and *garrafãos* (carboys) of copaiba oil that Pinto planned to deliver downriver, as well as his remaining household items. As he had done with other trade goods, Morris made notes on the production of copaiba oil:

> The tree producing this oil (*Copaifera Officinalis)* in size vies with the castanha, having a tall branchless trunk and a heavy crown of dark green foliage. It is found throughout the lower Amazonian valley and on the Tapajós, where it occurs in great quantities. It is frequently interspersed in the rubber forests. The oil, with which the whole tree is surcharged, is obtained by cutting large hollows in the trunk of the tree, which soon become full of a liquid from the top of which the oil is skimmed. It has an offensive and very peculiar smell, but is when fresh not nearly as bad as when it becomes old, the state in which it is usually seen. It forms an extensive article of commerce on the Amazon. We have often met with these trees in the forest rent asunder and shivered into many fragments and appearing as if struck by lightning. This however is not the case, and the Indians have told us that the trees burst themselves, probably from becoming overcharged with oil, and thus perhaps the loud explosions so often heard in the forests may be accounted for.[8]

Pinto, Morris and twelve Tapuya paddlers took positions in one canoe; Antonique and his Munduruku crew were in the other. For the next several days they made great progress down the river by sticking to the swiftly moving main channel. They passed the cachoeira of Acará almost without realizing it, so disguised were its rocks. Soon they came upon Bacabal, and Pinto opted to pass by it rather than encounter its zealous protector, Frei Pelino; and Morris offered no objection. When they camped on the shore that night, Morris once again gave his dried heads another smoking.

Pinto feared the descent through the cachoeira of Apuí more than any other obstacle. Morris described what happened in the most thrilling prose found anywhere in his writings:

> The crews of both canoes paddled vigorously, but that of Antonique was in advance when soon after we entered the rapid of Apuí. Just at the head of this rapid the wife of Cândido Pinto opened a small trunk and took therefrom a small china doll gaudily dressed and spangled, which had been blessed by the priest and represented some saint. This she laid on the seat of the canoe, and kneeling in the bottom of the boat clasped her hands and prayed to it to ensure us a safe passage. Before us we saw the canoe of Antonique tossing up and down fearfully on the foaming waters, and seeming every second as if it would be engulfed. The channel through which we were rushing was only about fifty feet wide. On one side was a low island, on the other a high bank, both heavily wooded, so that we seemed entirely shut in. Before us rose what appeared to be a wall of foaming water, seemingly six feet high, through which it appeared impossible for any canoe to pass. This wall rose and fell in great masses of seething water, and the roar was frightful.
>
> From the moment we entered the rapid Cândido Pinto sat with clasped hands exclaiming "Ah meu Jesu! Ah meu Jesu!" and we, sitting by his side, clasped instinctively the side of the canoe, expecting every second to find ourselves in the raging waters. The steersman stood firm as a rock, braced against his great paddle, not a muscle of his face changing, the man at the prow with his large paddle aiding him with all his strength, and the crew with their smaller paddles, seemingly using superhuman exertion. We saw the canoe of Antonique enter the wall of water and disappear! There was no time for a word. Before we could have uttered a syllable, we were in the midst of it ourselves. It was a moment of whirl and tumult; which memory can recall but which no words

> can describe. The time of suspense, though it was but a few seconds, seemed interminable.
>
> The water poured into the canoe. Before us we saw the great hill of Quatá seemingly close upon us, and the rushing waters at its base, in which we knew no loaded canoe could live for an instant, and it seemed inevitable that we should be swept into them. An instant later and the wall of water was past; the canoe swerved violently from its onward course towards the easterly bank, and with a shock which would have broken an ordinary canoe into fragments, the prow struck upon a sandy beach. All this happened in far less time than one can read the description. Words are all too weak to describe such experiences, and must ever feebly express the sensations of such moments.
>
> We found Antonique at the landing place. His canoe had come down without shipping any water, while ours from three large waves which had broken over us was more than half-full; and the gunwale, as she lay in the port, was only four inches above the level of the water. We had had a narrow escape, as even the old pilot confessed. Much to our surprise, Cândido Pinto laid all the blame of the peril on us, charging that but for the human heads, which were in the canoe, we, too, should have come through as safely as Antonique.[9]

The passage through the remaining cachoeiras was uneventful in comparison to Apuí. They arrived at Leverger's as the sun was setting. It had taken Morris little more than two weeks to return from the villages of the Campineiros, whereas it had taken him nearly three months to reach them. He arrived at Leverger's with no shoes, no hat, patched trousers, and shirtless—all these items had been lost or rendered useless during his journey. He was pale, emaciated, weak, and unkempt, with long hair and a beard. Leverger's little island plantation now seemed to him to be the lap of luxury.

As Ernest Morris recuperated, Leverger informed him that rumors had preceded Morris claiming that he was bringing thirty contraband heads down the river. In his written account, Morris stated that this was the first time he heard that trade in the heads was illegal. Plainly he was dissembling, and offered this as a weak defense to his subsequent actions. He smoked the heads one more time, then packed them in a plain crate which he entrusted to guarana traders on their way down to Santarem in canoes. He said goodbye to Leverger at Itaituba, then boarded the steamer bound for Santarem. Morris suspected that the steamer would be searched, so sending the heads downriver by canoe was an effective subterfuge. Upon arrival at Santarem, Morris was told to disembark and report to the *delegado,* the police commissioner. Morris ignored the summons and dispatched a command for the crate to be rowed over to the steamer. Once the box was aboard, after many nervous minutes, the whistle blew and the steamer departed from Santarem, carrying Morris and his heads. In return for turning a blind eye to the smuggling, Morris rewarded the steamer captain, a German named Hoeffner, with one of the treasured heads.

Ernest Morris's family in Indiana had not heard from him since early February 1877. When he arrived in Pará from Santarem in early May, he wasted little time in booking passage to New York; pausing just long enough to buy a cheap, used suit. He did not bother to write ahead, since he would beat any mail packets arriving in the United States. He set foot in Brooklyn in the last week of May, sent a telegraph to Indianapolis, and promptly contacted the Long Island Historical Society's Natural History division, requesting an opportunity to show the items he had brought back at their monthly meeting. His brief talk and display of the dried heads created such interest that he was invited to come back to give a lengthier description of his travels among the Munduruku.

The Long Island Historical Society's Natural History division was America's closest equivalent to Britain's Royal Geographical Society—an organization devoted to supporting scientific exploration and fieldwork. Morris's extended lecture took place on June 7, 1877, where the rooms were described as "overflowing with members of the society and their friends," including "many gentlemen and ladies of the first social and intellectual prominence in the city."[10] Morris delivered a talk light on academic details, but full of the adventure of his journey up the Tapajós and Rio Cururu, among the Campineiros, and the return down the churning river. He spoke without notes, in an enthusiastic, conversational manner. The *New York World* estimated that the artifacts he brought back, including the dried heads, were worth between $2000 and $3000.

When describing the episode when he was deserted by Antonique on the upper Rio Cururu, Morris let slip a detail: Antonique had wanted to take Morris's store of trading trinkets to go get the heads, and Morris only dissuaded him by pointing his gun at him. Whether Morris was embellishing the story at that point; or whether his mistrust of Antonique was that severe--or even deserved--must remain a mystery. What seems fair to conclude is that Morris and Antonique misunderstood each other.

Morris became especially animated when describing the descent through the rapids of Apuí, so much so that the recounting of the tale forced him to collapse in a chair and cover his face with his hands.

ERNEST MORRIS, THE SOUTH AMERICAN EXPLORER.

Ernest Morris, the South American Explorer. From: Frank Leslie's Illustrated Newspaper, August 18, 1877, p.409. From the collection of the author

After the lecture, Morris entertained individuals who wished a closer look at his artifacts. The *Brooklyn Daily Eagle* noted, "Mr. Morris thinks more of these Indian heads than anything else, and last night he fondled them as a mother would her child. They are, indeed, rare curiosities."[11] It is unknown whether the *Eagle* scribe was referring only to the heads, or was also including Morris himself as a curiosity.

The editors of the *New York World,* in a piece headlined "A Progressive American Boy," took up the theme that Morris himself was more interesting than his artifacts:

> We should be sorry to help spoil him by premature commendations, but his is not an unfavorable type of American character—a peculiar and interesting manifestation of that strange and vigorous manhood which is developing out of the many diverse elements of our nationality...It would be rather rash to say that Ernest Morris has chosen the wisest course for his own intellectual development, or that he is likely to do much, if anything, to advance the cause of science. But his own character, as shown in his brief career, is more curious than any specimen he can collect, and certainly not less worthy of study than the habits of the Mundurucus. They may teach us something of the man of the past. He is a remarkable example of the irrepressible boy of America, out of whom is to be evolved the man of the future."[12]

In these comments, the *New York World* was suggesting a startling idea: that exploration was not merely for the gain of human knowledge, but that adventure itself had its own moral imperative; and consequently, that the explorer was, at times, more interesting than his discoveries. One can easily imagine these words being read by nineteen-year-old Theodore Roosevelt, just home from his first year at Harvard. It may be no coincidence that a fortnight after Morris gave his talks, Roosevelt took off with a college friend to catalog the bird species of the Adirondack wilderness.

---

[1] An example of this statement can be found in: "Brazilian Wonders," *Indianapolis State Sentinel,* June 13, 1877, p.6.
[2] Morris, *In Search of Human Heads*, Chapter XII.
[3] Morris, *In Search of Human Heads*, Chapter IX.
[4] The name Morris uses for this plant is not found in any other source; but phonetically is like *Mucunã de caroço; or carcica paypaya (Carica pawpaw)*
[5] Whirlpools
[6] Morris, *In Search of Human Heads*, Chapter XI.

---

[7] Morris, *In Search of Human Heads,* Chapter XIII.
[8] Morris, *In Search of Human Heads,* Chapter VII.
[9] Morris, *In Search of Human Heads,* Chapter XIII.
[10] "Ernest Morris's Lecture," *New York World,* June 8, 1877, p.5.
[11] "Adventure," *Brooklyn Daily Eagle,* June 8, 1877.
[12] "A Progressive American Boy," *Rocky Mountain News,* June 9, 1877. A reprint of an article from the *New York World.*

6 YANKEE ORCHIDO-MANIA

After returning from his second journey to Brazil, Ernest Morris remained in Brooklyn for two weeks. Rather than selling any of the Munduruku heads immediately, he decided to take all ten of them back to Indianapolis for display at a series of lectures in his home state. The first lecture he gave was on June 26, 1877 to the Indiana College of Physicians and Surgeons—the most prominent group of men with scientific training in Indianapolis. At the time, the study of anthropology had an emphasis on a biology and race, so this audience of medical scholars would have included those with an interest in natural history and the study of mankind.

Morris followed up this private lecture with a public talk on July 6, 1877 at the Indianapolis Masonic Hall, billed as "The Upper Tapajós, and How I Obtained Human Heads." Admission was 50¢, out of which Morris received a percentage, minus the expense of renting the hall. With the proceeds from this and other engagements, he hoped to raise the capital needed for a new expedition, although he was undecided on the destination. At the Masonic Hall, he mentioned that he was considering a journey to Zanzibar, which six years earlier had been the African starting point of Henry Morton Stanley's expedition to find Dr. David Livingstone. Not coincidentally, at the time of Morris's lecture, Stanley was in the African interior mapping the route of the Congo River. It may be that Morris dreamed of teaming up with Stanley; or of surpassing Stanley's accomplishments. However, had Morris known of the ordeals of Stanley's current expedition, he might have changed his plans on the spot: Stanley emerged from the Congo a month later, having lost half of his 228 native bearers.

Morris gave several more talks during July in communities surrounding Indianapolis: Martinsville, Rushville, Spiceland, Franklin, Crawfordsville, and other Indiana towns. In Lafayette, the hotel owner claimed that the room in which Morris had stayed with his heads was now haunted. Morris's crowds were often small, netting him sometimes as little as $37.00. In one case, he had to cancel his talk when advance sales were so weak that he could not even meet the hall fee. In Rushville, Morris spoke in front of a room only one-third full; a sudden bout of illness forced him to halt his presentation. Afterward, a doctor attended him for several hours. The local newspaper explained that "his nervous system was overtaxed," but the truth of the matter was more likely a recurrence of malaria contracted on the Tapajós. Returning to Indianapolis, Morris gave his talk at the Meridian Street Church; his account enraptured a local fourteen-year-old girl, Catherine R. "Kate" Perry, the daughter of a successful Indianapolis grocery wholesaler. Ernest exchanged some words with the girl, and as a result they established a correspondence.

Morris's ambitions for a wider lecture circuit were put to a halt by the Great Railroad Strike of 1877, which paralyzed transportation throughout the United States during the months of July and August. These protests were so disruptive to commerce that they were only stopped after the intervention of state militias and federal troops. Morris lost some of the advance hall rental bookings deposits he had made; and between those loses and the profit he hoped to make, he calculated that he was out $600 due to the "Great Upheaval." Even so, the returns from his lecture tour gave him barely enough to outfit himself for another trip to Brazil, but Zanzibar was out of the question. Impatient to return to the tropics, he made plans to head to New York, where he intended to sell some or all the heads; and then purchase better outfitting than he had taken on his earlier trips.

Though his lecture tour was disappointing, Morris's pride must have been soothed when the August 18, 1877 edition of the national news magazine, *Frank Leslie's Illustrated Newspaper,* was published with profile article and a large engraving of Ernest posing with his human heads. The caption read, "Ernest Morris, the South American Explorer." This was enough of a boost to propel him on another round of lectures around Southern Indiana, though his audiences were no bigger than before. On September 6, 1877, the *Indianapolis Journal* announced that Ernest would be leaving the next week for his third excursion to the Amazon, though he had only a few more dollars than when he started out previously. The *Journal* made a last appeal to the citizens of Indianapolis to contribute to his efforts.

The issue of *Frank Leslie's Illustrated Newspaper* hit the streets just in time to have been seen by none other than Herbert Huntington Smith, the young Cornell naturalist and geologist. Smith had returned from Rio de Janeiro to the port of Baltimore on August 20, 1877, and was back in Manlius, New York by early September. He had toiled for the past two years assisting his mentor, Professor Charles F. Hartt, with fieldwork for the new Geological Survey of Brazil. The highlight of those efforts was noted in the world's foremost scientific journal, *Nature,* in December, 1876. Hartt, Smith, and other assistants had mapped Devonian era rock beds, composed of Oriskany sandstone and containing fossils, 1,000 feet lower than previously found. Above it they found a Carboniferous bed, completing a link to the layers of more modern periods above.[1]

One can only imagine the reaction of Herbert Huntington Smith when he learned that Ernest Morris--the adolescent, amateur butterfly-chaser he had shunned in Santarem--was now lionized in the American press as a "South American Explorer." He must have doubted that Morris ascended the upper Tapajós at all. Moreover, Smith would have realized that Morris's fame was solely based on a criminal act: smuggling contraband out of Brazil. In doing so—and then bragging about it through the press--Morris placed at risk all the work being conducted by credentialed American scientists in that country. Did Morris not realize that Brazilian officials had access to American newspapers? Could he not guess that there would be official protests made to the United States government? Smith pondered long and hard on how he could counter the sensational, P. T. Barnum-like affront that Morris had perpetrated.

Smith would have been further galled by the October 11, 1877 edition of *Forest and Stream,* the nation's bestselling outdoor activity magazine. Morris was now in Brooklyn, making himself available to reporters and attempting to raise more money through the sale of the dried heads. In addition to hailing Morris "...who, by his zeal and intrepidity, is fast winning the admiration of the whole country," the editors of *Forest and Stream* disclosed the goals of Morris's next voyage--which Smith and anyone else with knowledge of the Amazon would have recognized as beyond audacious, and foolhardy in the extreme:

> ...Having made his way up the Amazon, [he] will ascend the Xingu, one of its largest tributaries, which is utterly unknown beyond the accounts of Indians and half-breed traders...he intends to organize a large party of Mundurucu Indians to carry his boats past the rapids, and beat back the hostile natives if they should attempt to oppose his passage.[2]

After reaching the headwaters of the Xingu, which Morris believed were only a few days' march from the source of the Rio Cururu, he and his Munduruku crew would cross land over to the villages of the Campineiros. He would then descend the Rio Cururu, and from there continue up the Tapajós to the San Manuel river; and ascend that until reaching a point where he could portage to the Parana river and take that all the way to the port of Montevideo in Paraguay.

Morris's geographic guesses were woefully wrong: the headwaters of the Xingu are far south of the source of the Rio Cururu; and there is a vast distance between the upper reaches of the Xingu and any waterway that would lead to the Parana river. If Morris ascended the Tocatins river instead of the more westward Xingu, he could then follow one of its tributaries, the Araguaia, hundreds of miles to the south, and with luck connect to the Rio Apore, which then flows into the Parana; but Morris never showed any interest in the Tocatins, dismissing it as too unhealthy (malaria-prone) to traverse. Even had his guesses not been wrong, Morris was probably underestimating the cost of obtaining a crew of Munduruku willing to travel hundreds of miles from their territory and wage battles against Xingu tribes in the heart of the latter's homelands.

Pushing in to the upper Xingu was a bad idea that countless explorers who came after Morris attempted, costing them their lives. The most famous of these was Colonel Percy Harrison Fawcett, an experienced English surveyor of the Amazon, who in 1925 set out for the Xingu headwaters over land from the south. He disappeared, causing much publicity and speculation. Over the following decades, as many as 100 other would-be explorers—some of them delusional about their own capabilities—attempted to discover Fawcett's fate and died in the process. They succumbed to hostile tribes; starvation; tropical disease, festering wounds, and animal attacks. This, then, was the goal that Morris hoped to reach, forty-eight years prior to Fawcett.

It was a busy October for Morris: in addition to *Forest and Stream,* Morris's new plans were published in the *New York World* and the *Boston Daily Advertiser.* All his dried heads were purchased by Elias Lewis, Jr., of the Long Island Historical Society; suggesting that this deal was pre-arranged from the very beginning, and that Lewis was the one who had urged Morris to acquire the artifacts.[3] Morris also wrangled a sponsorship from a New York knife and axe company, adding their wares to his outfit to serve as objects of trade with natives.

The news article on Morris's next trip that appeared in the *Boston Daily Advertiser* turned out to be a pivotal event. It was through this Boston publicity that Ernest Morris's travel plans came to the attention of Edward Sprague Rand, America's foremost authority on orchids. Rand, then forty-three years old, was the scion of a prominent Boston family and a lawyer by profession, following his father's footsteps. He was a graduate of Harvard (1855) and Cambridge Law School (1857). He was also a writer: in 1859, his book of poetry, *Life Memories* was published. In 1875, one of Rand's poems was published in the anthology *Laurel Leaves*, alongside such luminaries as Henry Wadsworth Longfellow, William Cullen Bryant, John Greenleaf Whittier, and Oliver Wendell Holmes. It is worthwhile to recite that poem, "When My Ship Comes In," if only to contrast its sentiment to what was later disclosed about Rand's behavior at the time he wrote it:

'Twas years ago, full many a day.
Since my gallant ship went out to sea,
And summers and winters have lapsed away,
But still she never comes back to me.
Patient I wait, while others twine
Bright flowery garlands, or honors win.
And muse on the treasure that shall be mine
On the happy day when my ship comes in.

Oh, fierce the winds have lashed the tide,
And the waves have drank the pouring rain,
And towering icebergs yearly glide
To southern seas from the frozen main.
I know no fear, though upon the sea
The winds course free in their fearful sport.
For something ever assureth me
That my ship shall come to the homeward port.

When the storm rides wild on the icy air,
And bright 'neath the andirons the embers glow,
I sit and muse, and I little care
For the raging wind and the blinding snow.
On summer eves when the stars stoop low,
As if in love with the flowery lea,
Afar mid the sleeping trees I go,
And glorious visions unfold for me.

For summer and winter are one to me.

And the day is bright, be it storm or shine,
For far away, o'er a sunny sea.
Sails a treasure vessel, and all is mine.
I see the ripples that fall away
As she cleaves the azure waves before.
And nearer, nearer, day by day,
Draws the happy hour when she comes to shore.

" But what if she never comes," you say.
"If you never the honor, the treasure, gain?
It has made me happier day by day,
It has eased full many an aching pain.
It has kept the spirit from envy free.
Has dulled the ear to the world's rude din.
Oh, best of blessings it's been to me,
To look for the hour when my ship comes in. [4]

In these lines, Rand expresses that anticipation and hope for a better fortune can be as satisfying as actually coming into possession of such treasures. The irony is that Rand, at the time he penned this, was busy acquiring and surrounding himself with treasures—with little concern for how they would later be paid. In other words, he was not waiting for his ship to come in at all.

Edward Sprague Rand, circa 1870s. From an album of members of the Massachusetts Horticultural Society. Photo courtesy of the Massachusetts Horticultural Society.

Edward Sprague Rand Jr.'s true passion was horticulture. At age 21, he joined his father as a member of the Massachusetts Horticultural Society, and eventually was recognized as a lifetime member. During the 1870s, Rand amassed impressive plant collections in the greenhouse of his country estate, Glen Ridge, located in Dedham, Massachusetts. Rand's greenhouse was described in an article of *The Horticulturist*:

> The whole length of the greenhouse is about 135 feet, and its width twenty feet. It is divided into five compartments, namely: a rose-pit, forty-five feet by twenty; a conservatory, twenty-five feet by twelve; an orchid house, twenty-five feet by twenty; and geranium and camellia houses, forty feet by twenty; the first and last named being at the extremes, leaving the orchid house and conservatory in the middle, where the heat is greatest; a very neat and glazed potting-room being at the side of the conservatory, near the boiler. Good light and ventilation, so essential to orchids, are provided, and the experiment of ground glass, of which a portion of the roof is made, has proved entirely successful...the construction of the whole building is excellent.[5]

Throughout the 1860s and 1870s Rand penned a series of very popular gardening titles: *Garden Flowers, How to Cultivate Them*; *Flowers for the Parlor and Garden*; *Popular Flowers*; *Bulbs, a Treatise*; and others. Each of his titles was reprinted multiple times, by leading American publishers. Now in the public domain, Rand's gardening books are freely available on the Internet. In 1876, a year before encountering Ernest Morris, his tome *Orchids, a Description of the Species Grown at Glen Ridge* was touted as the first authoritative American work about orchid culture.

A chapter of Rand's orchid book gives the history of orchid growing in the United States, related by one of its earliest practitioners. It is fascinating to look at the first horticulturists who adopted this interest—not unlike the concept of contact tracing in following the spread of an infectious disease. Rand identifies the first collection of orchids in the United States as belong to Boston merchant John Wright Boott. In 1838, not long after several greenhouses in England had imported and cultivated orchids, James Boott sent plants from London to his brother John in Boston. For John Wright Boott, the orchids were a recreation to distract him from the burdens of business. Boott received more specimens before his death in 1842, and bequeathed all of them to John Amory Lowell, a notable Massachusetts businessman and philanthropist whose lawyer father had been one of the founders of the Massachusetts Horticultural

Society.

Lowell was the first to exhibit orchids in the United States, at a meeting of the Massachusetts Horticultural Society. Lowell moved from his country residence in 1853, selling most his collection to Edward Sprague Rand. Rand kept his orchids for nine years, importing additional species from England. During the 1850s, Rand had the finest collection in America. He eventually donated his plants to Harvard in 1865, where they were added the herbarium of the Botanic Gardens in Cambridge. In the late 1860s, Rand relocated to an estate called Glen Ridge in Dedham.[6] Originally just ten acres in size, Rand greatly expanded the property. In 1869, a report of the Massachusetts Horticultural Society described the grounds:

> ... with great diversity of surface, high and low, broken and dry, with no propitious line for avenue or approach. It was appropriately named 'Glen Ridge.' Since that time, however, by successive additions, its area has been increased to nearly *two hundred acres*. Its location is about one mile from Dedham village, and nine miles from Boston. On the North, it is flanked by Charles River, which here expands almost to a lake, the shore and estate presenting a pleasing variety of the picturesque and beautiful--a combination of hill-side and valley, ridge and dell, wild-wood and forest, rills and runs, pines and pepperidges [sic], oaks, beeches and birches--giving ample material and space enough to employ the busiest brain for years to come. That portion of the property which first strikes the eye, consists of the old 'Fair Grounds' of the Norfolk Agricultural Society, which only two years ago, before Mr. Rand's purchase, presented an uninteresting aspect, destitute of any great attractions, yet essential for access to the principal estate and affording wings and supports to the same.

The passion for orchids spread from Rand to others within the Massachusetts Horticultural Society, and to other growers, such as James Fisk Allen and Frederick Lothrop Ames, a railroad tycoon--the richest man in Massachusetts. Rand had sold much of his first orchid collection to Ames. Gatherings of the Massachusetts Horticultural Society also attracted a dedicated florist from Albany, New York: Louis Menand. Menand's knowledge and enthusiasm as a floriculturist was shared by two other wealthy Albany residents, Gen. John F. Rathbone and Erastus Corning Jr.

Edward S. Rand's first collection of the 1850s outshined several other collectors of that decade. Starting in 1854, the exhibitions of other horticultural societies in New York, Brooklyn, and Maryland included mentions of orchids. One of the first New York growers was horticulturist Thomas Hogg; Hogg's father, before emigrating to America, had been the gardener to William Kent, a landscape architect who had the largest private garden in England. When Hogg's orchid collection was disbursed in 1854, several other enthusiasts used those plants to start their own collections: Dr. James Knight, Cornelius Van Voorst, Jesse Paulmerre, and Isaac Buchanan. In the mid-1850s, gardener George Hamlyn won several prizes for orchids on behalf of his patron, Brooklyn broker William Clark Langley.

Over the years, orchid cultivation grew to be the focus of Edward S. Rand's horticultural interest at Glen Ridge. It was an expensive hobby, requiring special climate-controlled greenhouses, substantial shipping costs, and constant attention. After hearing about Morris's journeys, Rand was convinced that he was the ideal person to help him to acquire new orchid specimens from one of the world's best orchid habitats, the Amazon basin. For over twenty years, Rand and other American orchid growers had been dependent on two English nurseries: James Veitch & Sons and Hugh Low & Sons.

Where and when Rand and Morris first met is unknown, but Rand's excitement about the collecting project soon escalated to the point where he declared that he would accompany Morris to Brazil and fulfill his lifelong dream to study tropical flora in their natural setting. Moreover, Rand admitted that he was in poor health and that his doctor had recommended a change in climate to benefit his ailments.

That, at least, is the impression that Rand gave to Morris at the onset of their relationship. It may not have occurred to Morris that it was odd that Rand, a prominent lawyer with a wife and four children back in Boston, could forego all his commitments and sail off to Brazil at the drop of a hat. It was so impulsive a move that Morris and Rand went together to a New York notary public on October 23, 1877, as signed as witnesses on each other's passport applications. Ten days later they boarded the schooner *Jacob E. Ridgway,* captained by John C. Townsend, bound for Pará, Brazil.[7]

At some point, Rand must have revealed the truth to Morris: his passion for horticulture had driven him to overextend his finances to fund his greenhouses; moreover, as a lawyer trusted with handling the estates of others—including his own relatives—he had embezzled large sums and poured all of it into his hobby. His properties were over-mortgaged, worth much less than what he had paid for them. Earlier that year, he had been jailed; and had only been released after selling his Dedham estate and taking the Poor Debtor's Oath, i.e. swearing that he had no other hidden assets. He either deserted his family or they left him; and all he had left in the world was in the luggage he carried on the schooner—which was small enough, but probably more than he had admitted to the courts. He was soon disbarred, and likely would have faced imprisonment if he returned to Boston. The Boston newspapers had reported Rand's troubles with glee, as did other newspapers all over the United States. The *Boston Daily Advertiser*'s account reveled in the scandal using melodramatic prose:

> Rand's Downfall: Lessons from the Drama of Life: Unfortunate Results of Mistaken Ambition: The Old Story of Extravagance and Wrong
>
> The extraordinary process of *ne exeat republica* has not been resorted to for many years in this Commonwealth until the present week, when the parties interested deemed it necessary to apply its provisions to the person of Mr. Edward S. Rand Jr., whose general reputation as a barrister and gentleman has been excelled by but few in Massachusetts. The very name of the proceeding indicates its purport—to restrain the respondent from going out of the country—and thus we see a brilliant, genial cultured man of letters, learned in legal lore, filled with a love of the beautiful in nature, inspired with no mean poetic genius, connected with some of the most aristocratic and historic families of old Massachusetts, estopped by a mandate, under seal of its Supreme Judicial Court, from going out beyond the boundaries of his native land, for fear that justice will not be satisfied in regard to some trusts betrayed, some debts due, some funds endangers, or some accounts unadjusted...
>
> ...His residence and its surroundings showed the very poetry of Life. His estate was in a beautiful location in the upper portion of the village of Dedham, the locality being known as "Connecticut Corner," there were between 100 and 200 acres, in which nature had done much to gratify the aesthetic tastes of this ardent student of the harmonics in nature and art. Forest pines overshadowed some of the walks and drives, which, winding in tortuous labyrinths amid flowers, ferns, rare plants and fine shrubs, environed and traversed this garden of paradise; while others were shaded by young trees, whose symmetry was made a

> constant care, and their harmony of foliage a study and a pride. Statuettes and fountains lent variety and beauty in a wilderness of flowers, whose glory of rhododendrons were the wonder of the visitor, while orchids from South America dazzled the eye of the beholder. His green houses were stored with the choicest specimens of floriculture, and in such profusion and rarity that the cost of the maintenance alone is guessed often as above $10,000 yearly. The mansion of 30 or 40 apartments, included elegantly furnished boudoirs, banquet rooms, parlors, library, and all the accessories of comfort and pleasure in prodigal array; while the wine cellar and side board were supplied from those vintages which a connoisseur only selects. Horses and carriages were always kept in plenty, while man-servant and maid-servant thronged the place to do the bidding of master and mistress. From piazza and window, from turret and hall, the undulations and various beauties of this product of nature and the gardner's skill filled the eye; while from the summit of a stand-pipe, erected at great cost to supply the place with water from the Charles River, landscape and village for miles around could be seen on every hand. The residence was fine ten years ago and has constantly progressed in improvement. Rich paintings, elegant carpets, exquisite statuettes, a wealth of beautiful ornaments, and a library filled with well-selected volumes gave enchantment to the other furnishings of the above. In summer his cheeks were fanned at his own door by the breezes which swept down from the hills at Mount Desert[8]; while at home a special pride was the elegant dejeuner or the superb dinner party, where he was master of ceremonies, welcoming guests whose society a prince might covet. In his business, none were more eagerly sought for advice upon titles, conveyances and real estate transactions generally; while he was known as a most methodical and indefatigable worker. This was the man. On Tuesday he testified, in answer to counsel, that his earthly possessions were *only the clothes he wore.*[9]

Rand likely presented Morris with a less-embellished version of his woes, and did indeed not look healthy. But whatever other illnesses Rand suffered from, his worst disease was orchidelirium, the craze for collecting orchids; and he was not the only victim of this affliction that Ernest Morris would encounter.

---

[2] "Ernest Morris," *Forest and Stream,* November 8, 1877, p.272.

[3] "Morris," *Brooklyn Daily Eagle*, November 4, 1877.
[4] Gill, William Fearing, Henry Wadsworth Longfellow, and Louisa May Alcott. 1879. *Laurel leaves: original poems, stories, and essays.* New York: R. Worthington.
[5] "A Beautiful Conservatory," *The Horticulturalist and Journal of Rural Art and Rural Taste*, 1875, v.30, p. 350.
[6] Rand, Edward Sprague. 1888. *Orchids; a description of the species and varieties grown at Glen Ridge, near Boston, with lists and descriptions of other desirable kinds: preface by chapters on the culture, propagation, collection, and hybridization of orchids; the construction and management of orchid houses; a glossary of botanical terms and significance of their names; the whole forming a complete manual of orchid culture.* Boston: Houghton, Mifflin and Co. http://catalog.hathitrust.org/api/volumes/oclc/19421590.html
[7] "A Remarkable Man," *Brooklyn Daily Eagle*, November 6, 1877.
[8] In 1868, Rand had purchased a cottage at Bar Harbor, Mount Desert Island, Maine, near the current Shore Path walking trail.
[9] "Rand's Downfall," *Boston Daily Advertiser*, April 19, 1877.

Chondrodendron tomentosum (curare). From: Bentley, Robert. *Medicinal plants. Being descriptions with original figures of the principal plants employed in medicine and an account of the characters, properties, and uses of their parts and products of medicinal value.* London: J & A Churchill, 1880. Image courtesy of Biodiversity Heritage Library. http://www.biodiversitylibrary.org

Victoria Regia. From: Allen, John Fisk. *Victoria regia, or, The great water lily of America: with a brief account of its discovery and introduction into cultivation.* Boston: Dutton & Wentworth, 1854. Image courtesy of Biodiversity Heritage Library. http://www.biodiversitylibrary.org

Araras by Ernesto Lohse. From: Göldi, Emil August. *Album de aves amazonicas.* Belem, Brazil Museu Goeldi de historia natural e ethnographia, 1900-1906. Image courtesy of Biodiversity Heritage Library. http://www.biodiversitylibrary.org

Cattleya eldorado splendens by John Nugent Fitch. From: Orchid album: comprising coloured figures and descriptions of new, rare and beautiful orchidaceous plants. London :B. S. Williams,1882-97. Image courtesy of Biodiversity Heritage Library. http://www.biodiversitylibrary.org

Morpho helenor by W. H. Lizars. From: *The Naturalist's Library. Vol. XXXI. Entomology. Foreign Butterflies.* Edinburgh, W. H. Lizars; [etc., etc.]1833-. Image courtesy of Biodiversity Heritage Library. http://www.biodiversitylibrary.org

PLATE LV by Henry Walter Bates. From: Bates, Henry Walter. *Contributions to an insect fauna of the Amazon Valley : Lepidoptera : Heliconidae.* [s.l. :s.n.],1861. Image courtesy of Biodiversity Heritage Library. http://www.biodiversitylibrary.org

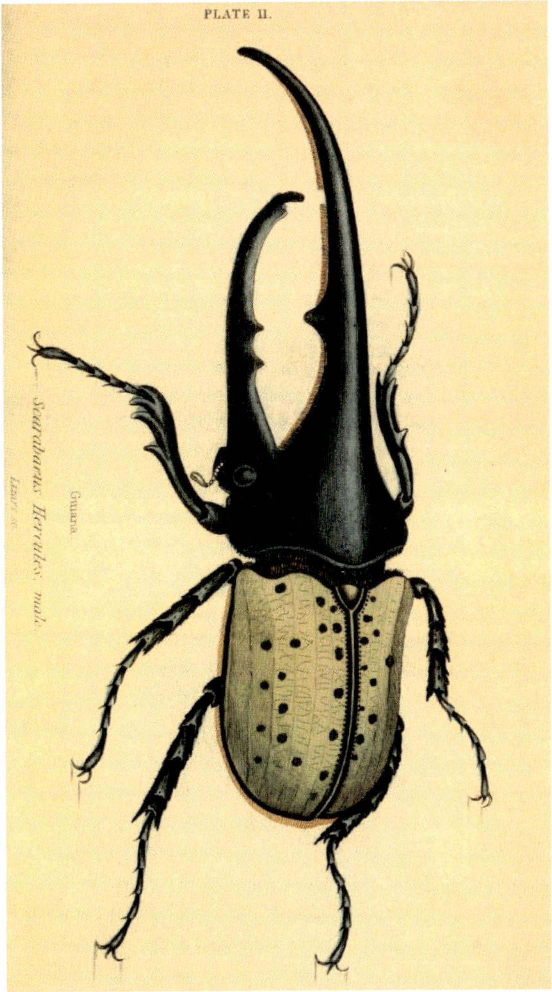

Scarabaeus hercules by W. H. Lizars. From: Duncan, James. *The natural history of beetles : illustrated by thirty-two plates, numerous wood-cuts, with memoir and portrait of Ray.* Edinburgh: Harry G. Bohm, 1852. Image courtesy of Biodiversity Heritage Library. http://www.biodiversitylibrary.org

Epidendrum atro-purpureum var. randianum by Henry George Moon. From: *Reichenbachia: Orchids illustrated and described.* St. Albans :London,New York,F. Sander & Co., Orchid growers & importers;H. Sotheran & Co.;I. Forstermann, [etc., etc.]1888 [-90]Image courtesy of Biodiversity Heritage Library. http://www.biodiversitylibrary.org

Paphinia cristata lindl var. randi by P. de Pannemaeker. From: Linden, Jean Jules. *Lindenia : iconographie des orchidées.* Gand [Belgium] :Impr. F. Meyer-van Loo,1885-1906.Image courtesy of Biodiversity Heritage Library. http://www.biodiversitylibrary.org

Ernest T. Morris, circa 1876. Photo courtesy of Marianne Lanman Salaymeh.

Ernest T. Morris lecture handbill, 1876. Courtesy of Marianne Lanman Salaymeh.

Ernest T. Morris, circa 1879. Photo courtesy of Marianne Lanman Salaymeh.

Kate Perry Morris, circa 1883. Photo courtesy of Marianne Lanman Salaymeh

Ernest T. Morris, circa 1881. Photo courtesy of Marianne Lanman Salaymeh.

Erastus Corning Jr. by Charles Loring Elliott (1812-1869). 1864, oil on canvas, 27 ½ H x 22 ½ W. Courtesy Albany Institute of History & Art; Gift of Mr. and Mrs. Francis W. Rawle, Jr., 1981.6

7 CALL OF THE XINGU

After arriving in Pará following a rough ocean journey lasting twenty-four days, Ernest T. Morris and Edward S. Rand boarded a steamer going 650 miles up the Amazon, to the town of Óbidos. Óbidos, home to 800 people, was the most populous settlement between Santarem and Manaus, and was the center of the cacao trade. To Morris it seemed a good place to hole up for a couple of months until the rainy season--just beginning in December--subsided. Morris had learned his lesson about traveling during the seasonal downpours. Óbidos was less than fifty miles downriver on the Amazon from Morris's favorite spot, Juruti; but afforded more comfort and commercial traffic that Rand desired. They arrived at Óbidos on December 9, 1877.

They set up residence in a mud and tile hut about a mile outside of town, near a lake called the Lago do Juncal. It had forests nearby, and many paths leading into the woods. However, Morris soon discovered that it offered limited variety for a specimen collector. The birds were common; there were few butterflies; and a small number of common orchids. Morris was surprised that the nature of the forests, the soil, and the fruit season differed so much from Santarem, which was only fifty miles downriver, but on the southern bank of the Amazon. The northern bank where he now resided seemed a much different country.

By late January 1878, Morris had lost all patience with the poor collecting opportunities that the Óbidos area provided, and convinced Rand to accompany him to go up and across the river to his favorite locale, Juruti. There, they found the nearby forests full of birds and insects, including butterflies: *Morpho hecuba*; *Morpho menelaus*; *Morpho helenor*, *Morpho rhetenor*, and others. Virtually cut off from communication with the outside world, Morris penned a long letter to the *Brooklyn Eagle* which concluded:

> Now we are undergoing the discomforts of the rainy season. The rain is heavy and incessant. There is an air of gloom over the landscape. The forests are partly hidden by the driving rain and the air is laden with an odor of decaying vegetation. I contrast the graceful palms of the delta with these grand gloomy trees. They seem to droop with a weight of moisture. The earth is reeking, and a steam of vapor fills the air. Now and then a bright day comes, and the naturalist can work. And on these bright days how the aspect of nature is changed! The leaves and the flowers shine in the tropical sun, and at night the stars and the clear heavens seem scarcely above the forests. More and more am I convinced that Brazil is the paradise of the naturalist.[1]

In mid-March, Morris hired a canoe and four older boys from Juruti to venture up the Amazon towards an outlet leading to a fresh, clearwater lake, the Lago Grande de Juruti. Morris told two different stories about what occurred on the lake: one version was recorded on the spot in his journal; and another, much modified version was told once he returned to the United States from Brazil. In his journal, he notes that on March 17, the canoe entered the Lago Grande de Juruti, which he described as a beautiful sheet of water, dark, but clear, surrounded by high hills and rolling land. It resembled a wide river more than a lake. They aimed for the old village of Juruti Velho, which had been mostly abandoned when residents moved downriver to the Amazon and founded the new village of Juruti. It lay about a mile south from where they had entered the lake.

As they were crossing to the village, a sudden storm erupted, creating waves that threatened to capsize the canoe. They took on water and had to bail furiously, but after twenty minutes reached the old village, where the few remaining residents had been watching them with concern. The sun was out by that time, so they spread out their clothes and baggage to dry, and made camp in an old abandoned shanty that lacked doors and windows. Despite the dilapidated buildings, Morris thought it a beautiful place, ideal for swimming, with sandy white beaches. He spent two days in the old village, and on one of those days went a little further up the river. By then the date was March 16. Morris's journal then has a gap, until March 24 when he scrawled: "Taken sick. No more traveling." Then, another gap and a later note: "Left suddenly for home, April 24; sailed from Pará, May 12, 1878."[2]

The implication from these entries is that Morris became ill after arriving at the old village of Juruti Velho. Due to that, he curtailed his excursion and returned to the newer village of Juruti, where he had left Rand. His was in poor condition for a month, save for a brief period, when on April 6, wrote to the *Indianapolis Journal* from Juruti:

> I have been awaiting the ending of the rainy season before starting into the interior. The rains are now nearly over, and I am anxious to be at work again. My health is not good. My outfit is miserable, but I shall do my best to make a thorough examination of the river Xingu, and will try to cross the country between the Xingu and the Tapajós. I go alone. If my health will permit I shall work harder than I did on the Tapajós, just to show you people that four years' pulling against the stream has not discouraged me. But it is hard work, when many are expecting you to accomplish much on nothing.[3]

Ernest Morris was not exaggerating that many people assumed great things from him. His scientific contacts back in Brooklyn had urged him to go up the Xingu, for they believed he might make great discoveries there.

On April 8, 1878 (long before any word of Morris's bad outcome on Lago Grande de Juruti would have reached the United States), the *Brooklyn Eagle* published a long article on the aims of Morris's projected expedition, throwing Morris right into the middle of the raging scientific debate of the day between the theories of monogenism and polygenism. The conflict was between those who believed in a single origin of mankind, which then dispersed over the globe through migration (monogenism); or multiple origins, which asserted that humans had developed separately on different continents (polygenism). It was an argument that had a great deal to do with belief in racial superiority (polygenists often argued for the primacy of Caucasians above all other races); but the *Eagle*'s anonymous author (Elias Lewis Jr. of the Long Island Historical Society would be a good guess), was a monogenist, and had a different take:

> It is even probable that the explorer in South America is traversing the birthplace of man...Mr. Morris is engaged in the study of the history of the human race. Deep below the mighty forests which cover that continent lie the relics of what was probably the first civilization the world has ever seen, beside which the necklace of Helen and the pottery of Cyprus are but the baubles of yesterday...The waste places of the Southern Continent

> conceal more testimony, and it is for intrepid and intelligent workers, like youthful Mr. Morris, to piece together the evidence which awaits discovery.[4]

Morris's Brooklyn sponsors, or at least a single prominent one, had charged him to go forth and find traces of a great undiscovered civilization. Decades later, explorer Percy Fawcett came to believe the same thing existed, and called it the Lost City of Z.

Burdened with such heavy expectations, Morris's sense of failure must have been enormous. By mid-April, his health was still bad, and he decided he had to return to the United States to convalesce. It is also likely that Morris came to realize that his dream of ascending the Xingu was an impossibility. Back in December, while in Óbidos, he learned that his former guide, the rubber trader Barnabé, had died at the mission at Bacabal. Morris had been counting on getting Barnabé's help to recruit a crew of Munduruku warriors. Morris must have realized that he would not be able to convince the Munduruku to make such a long, dangerous, journey; and that the reputation of the tribes on the Xingu was such that he could not consider making the trip alone.

As Morris neared New York City in early June 1878, he also must have calculated that abruptly returning home due to a tropical illness would discourage his supporters from backing any future expeditions he might make. He first business in New York was to sell John Akhurst, the Brooklyn collector, 5446 butterflies that he had brought back. When Morris sat down for his first interview, he had a dramatic tale of his misfortune on Lago Grande de Juruti that he related breathlessly to the *New York World*:

> There suddenly came down the lake an immense white-capped wave; it was as if a narrow gust of wind had caught up a strip of the water from bank to bank, and was carrying it like a wall down between the hills. Jumping to his feet with the assistance of the Indians he kept the canoe head on, and the wave passed beneath it, lifting them high and letting them down with a thump. But now, with almost miraculous quickness, the sky grew black with clouds and lightning flashed in every direction. "I never knew what a thunderstorm was before," said the young man, rising from his chair and holding his arms wide open to express the magnitude of the storm. "It was like looking at a driving hail-storm upon which the sun is shining. The Indians were frightened out of their wits. Volley after volley of thunder rattled among the great

> trunks of the trees in the forest. In an instant the lake was a frothy sea, and the next moment we were all in the water. Fortunately, we were not far from the land, and all being good swimmers, we succeeded in reaching a rocky point that jutted out into the lake. On that rock I lay for thirty-six hours delirious...It was six days before I was able to start for Juruti, where I finally arrived. I was now powerless to continue my exploration. I had nothing left but a fine collection of butterflies." Here the lad paused, and the reporter asked him what he intended to do next. "What do I intend to do?" he rejoined indignantly. "Why, to go back, of course, and to stay until I have explored every inch of Brazil. I shall go back to South America, and won't make another failure of it, you may depend!"[5]

In Morris's recasting of the events on Lago Grande de Juruti, his canoe capsized during the freak storm, losing all his gear, and the struggle to reach shore nearly killed him. He could not afford to admit his health succumbed to a tropical illness.

While Ernest T. Morris was engaged in his abbreviated expedition of 1877-1878, the other young American naturalist of the Amazon--Herbert Huntington Smith--devised a plan to publicize his own experience in that region, and to establish himself as the leading authority on travel in Brazil. Using his contacts at Cornell and his resume as a former assistant newspaper editor, Smith leveraged a writing contract with *Scribner's Monthly* magazine, the leading American literary periodical. *Scribner's* agreed to fund an excursion to Brazil to be made by Smith and a popular illustrator, James Wells Champney, in return for a series of travel narratives on the people, landscape, and natural history of Brazil. These articles would then be compiled and expanded to a full-length book. No account of life in Brazil had been written for general readers since the Agassizes' *A Journey in Brazil.* Brazil and the Amazon had changed greatly since the Agassizes had visited in the early 1860s; the rubber industry and steamer traffic had transformed travel throughout the country. Moreover, there was a renewed interest in the United States in Brazil, due to Emperor Dom Pedro II's warmly received tour of the United States in 1876.

Smith and Champney embarked from New York on July 6, 1878, aboard the new steamer, the *City of Rio de Janeiro,* on what was billed as "a tour of exploration and observation." It was reported that the Brazilian government would offer every assistance—evidence that Smith used the official contacts he gained while working for the Geological Survey of

Brazil. Champney left behind his wife, the writer Elizabeth Williams Champney, to look after their young children. Their destination was Pará; and from Pará, by steamer up the Amazon. They had every advantage: talent; knowledge; sponsorship; experience; equipment; and the support of the Empire of Brazil.

Meanwhile, after giving a few interviews, Ernest Morris--his health now reinvigorated—visited "a leading university" in Boston; probably Harvard, the alma mater of Edward S. Rand. He went to solicit their interest in sponsoring a collecting expedition to Brazil. He was informed that the institution would be pleased to accept any donations to their collections, but would not pay for that service. One can only imagine the chilled tone in which that message was delivered. Morris may have miscalculated by name-dropping Edward S. Rand, who was by then a pariah in Boston society. The Boston newspapers were far less interested in the "Boy Naturalist" than they were in the fugitive Brahmin. Ernest defiantly obliged their curiosity: "He [Rand] is very pleasantly located and has a splendid outlook for accumulating a fortune; and, mark my words, when he does obtain it, he will—as he has many and many a time remarked to me—return to Boston and pay all his debts, both principal and interest."[6] A week before Morris made these comments, Edward S. Rand had formally been disbarred from the legal profession.

If the reception he received in Boston discouraged Morris, he had little time to dwell on it, for good news awaited him back in New York. A wealthy beetle, butterfly, and orchid collector, Erastus Corning Jr., of Albany, was interested in outfitting Morris's next trip to the Amazon. Whether Corning had learned of Morris through newspaper accounts, or was introduced by a communication from his acquaintance, Edward S. Rand Jr., is not known, but whatever the case, Morris could not have conjured a more obsessed and profligate patron than Erastus Corning Jr.

Erastus Corning Jr. was the only surviving male heir of Erastus Corning Sr., who had built a financial empire based on his hardware business. Corning Sr.'s hardware holdings included all kinds of iron tools and fasteners, including large implements such as stoves and farming tools. He had the foresight to start his own iron foundry in Troy, New York, to manufacture railroad track and train parts at the onset of the rail boom in America, assuring his fortune. He then diversified into banking and land speculation; and later served as mayor of Albany, as a State senator, and as a member of Congress. At his death in 1872, Erastus Corning Sr. was one of the richest men in America.

His son, Erastus Corning Jr., born in 1827, attended Union College and then joined his father's business, eventually becoming president of the Albany & Rensselaer Iron and Steel Works. He also was president of the Albany City National Bank; and a director of the New York Central & Hudson River Railroad. Although active in politics as a Democrat, Erastus Jr. declined several opportunities to seek public office. Like his father, Erastus Jr. followed the Temperance movement, and liquor was not served when he entertained at his city and country houses. However, this was the only sense in which he might have been called restrained, because he pursued several pastimes with extravagance; was charitable to a fault; and loved to entertain others and show off his collections. He loaned money to others to fund their risky business ventures, and often never was reimbursed when those projects failed. He donated money to schools, and gifted the land for the Episcopal Cathedral of All Saints in Albany.

Though he kept a fine house on State Street in Albany, Erastus Corning Jr.'s pride was his country estate, located in the Glenmont/Kenwood area just south of Albany on the Hudson River. He called the grounds "Ta-wass-a-gun-shee" a name associated with the hill upon which Albany was built. Starting in his early twenties, Corning transformed the property into a model farm, breeding Jersey and Hereford cattle and Cochin ("Shanghai") fowls. He was a horse fancier, who for many years owned the champion trotter, "George Palmer;" He helped establish a race course for trotters at Island Park, within present-day Menands, New York. He adored dogs; he overpaid for expensive purebred puppies; and if visitors grew attached to the canines, Corning would give them away as gifts. Inside his large country house, Corning set up cases to display his beetle and butterfly collections.

At about the same time—the early 1850s—he purchased thirty orchids in England and brought them home. By the late 1870s, he had fourteen separate greenhouse structures on the property. No images of the conservatories have survived, but an 1886 horticultural journal offered details on the structures devoted to orchids:

> Nine houses are devoted to the orchids, to wit: a Cattleya house, 100x20 feet; Cypripdium house, 50x17 feet; Dendrobium house, 100x20 feet; Oncidium and Lycaste house, 60x22 feet; Phalaenopsis house, 50x22 feet; Vanda house, 40x17 feet; a north or cool house, 36x12 feet, devoted to Masdevallias, and another cool north house, 60x12 feet, for Odontoglossums. There is, in addition to these, a resting house, 30x17 feet, where deciduous orchids which need a long period of rest, during which they are

kept dry and cool, are housed.[7]

His inventory was cited as the most extensive orchid collection in the United States by no less an authority than Edward S. Rand Jr. Corning's head gardener, William Gray, was a scientific expert in his own right; and the assistant gardener, Frederick Goldring, had been trained at London's Kew Gardens as an orchid grower. By 1880, Corning's trove of orchids was said to be valued at $100,000.

Erastus Corning Jr. was an early, prime example of the intersection of wealth and orchid culture, but in truth, few active businessmen or financiers could afford the time and fanatical devotion required for care of these plants. Money for greenhouse operation was a prerequisite. The ranks of early orchid collectors included professional floriculturists; retired men of wealth; heirs to wealth; and (later) the wives of wealthy men. The majority of the above started out with a generalized passion for gardening that gradually narrowed to a focus on orchids. In terms of expense and compulsion, orchid growing rivaled another historical floriculture craze, tulipomania. Orchido-mania (a contemporary nineteenth century term for the passion, now often called "orchidelium") spread to America in the 1850s, having first taken root in England in the 1830s.

Unfortunately, no records exist that explain where and how Erastus Corning Jr. and Ernest Morris met; or their impressions of each other; or even the details of their business transactions. What is known is that Morris spent the next four years collecting orchids with Corning as his primary patron, and quickly learned what pleased him. While there is evidence that Morris delivered plants directly to Corning's "Ta-wass-a-gun-shee" estate, we do not know if Morris was invited to stay there as a guest, though considering Corning's universal acclamation for being a good-natured host, that seems likely.

In addition to gaining Corning as a client, Morris received other good news. The *New York World*, which had featured Ernest in several articles, commissioned him as a Special Correspondent to produce travel narratives of his next trip. The *New York World* had been owned for the previous two years by robber-baron Thomas A. Scott, president of the Pennsylvania Railroad. Scott was accused of using the newspaper to promote his business interests—a charge that would also be leveled at the *World's* next owner, Jay Gould. Morris was hired as a columnist in hopes that he might capture some of the same readership that had been enthralled by Henry Morton Stanley's accounts for the *New York Herald*.

To add to this abundance of good fortune, Morris received the following letter from a Philadelphia firm:

> Mr. Ernest Morris:
>
> Dear Sir—If you can so arrange your matters in this country as to enable you to join our party I will be glad to make you naturalist and interpreter of our party, now under way for northern and eastern Bolivia via the Amazon and Madeira rivers, on the terms we have discussed. Very truly yours,
>
> C. P. Mackie[8]

Charles P. Mackie was one of the principals of Mackie, Scott and Company, one of two American engineering firms contracted by a group headed by a visionary American civil engineer, Colonel Earl Church. Church's dream for over a decade had been to provide a way for Bolivia to send its goods to the Atlantic; and for Brazil to send rubber directly to the Pacific, for far less than it cost to ship goods past the southern tip of South America, Tierra del Fuego. Church mapped a route for a railroad to bypass the cataracts on the Madeira river that blocked water navigation to its tributary the Mamore river, which descended from mountains in the heart of Bolivia. Under Church's plan, only about 130 miles of rail would need to be laid through the forest to accomplish the link between navigable sections of the river.

By June of 1878, work on the Madeira-Mamore Railway was underway, and a few miles of track had already been laid. While a separate contractor, the P. & T. Collins Co., was laying the tracks, Mackie & Scott were charged with transporting men and materials to the terminus on the Madeira river; and were also given rights to develop all economic resources along the railroad's route. To determine what those resources might be, Mackie & Scott needed a naturalist and a geologist to explore the planned route of the railroad; they offered the former job to Ernest Morris.

Morris agreed immediately to the proposition. As a member of this party, he could now officially call himself a professional Naturalist—no longer could anyone call him an amateur. With these arrangements in hand, in late June 1878, he returned briefly to Indianapolis to visit his family and friends. One of those acquaintances, Jeremiah Johnson, a clerk in the U. S. Attorney's office in Indianapolis, pleaded with Morris to let him come along on his next trip, as did his nineteen-year-old younger brother, Frank Morris. Frank had heard Ernest mention that an enterprising young man could make a fortune as a rubber broker, and wanted to test that opportunity. The Morris brothers packed for the journey back to New York, while Johnson promised he meet up with them as soon as he could settle his job obligations. Before leaving, Ernest also bragged about another offer he had pending:

> I may write a book when I return this time; in fact, I expect to do so. I have already been offered $600 for my notes alone in New York, but I have declined the offer. I am laboring to make a thorough naturalist of myself. I think I have been cut out for that, and I expect to make it the aim of my life. I have no one to thank for what success I may have achieved so far, and I intend to keep working now that I have a start.[9]

The Morris brothers arrived in New York in July 1878. Johnson arrived a day or two later, but they had to wait to catch the August 6 departure of the new steamer *City of Pará* for Pará—they were a month behind the steamer that was carrying Herbert Huntington Smith to Brazil to gather material for his own book. *The City of Pará* made very good time, reaching St. Thomas in the Virgin Islands by August 12 and Pará seven days later, on August 19, 1878. The same journey had taken Ernest Morris twice as long on his previous trips, so he could be forgiven for hoping for good things from this expedition; but very little unfolded as he had plann

---

[1] "The American Tropics," *Brooklyn Daily Eagle*, April 7, 1878.
[2] Morris's handwritten journal for 1877-1878 resides in the collection of Ernest T. Morris papers held by descendant Marianne Lanman Salaymeh.
[3] "General Notes," *New York Times*, May 21, 1878, p.4.
[4] "Mr. Morris in South America," *Brooklyn Daily Eagle*, April 8, 1878.
[5] "Ernest Morris Again," *Indianapolis Journal*, June 19, 1878, p.7.
[6] "E. S. Rand Jr. in Brazil," *Boston Globe*, June 19, 1878.
[7] Beard, Edward L. 1886. "The progress of orchid culture in America: read before the Massachusetts Horticultural Society, March 6, 1886". *Transactions of the Massachusetts Horticultural Society*, 1886. p. 123-151.
[8] "Ernest Morris," *Indianapolis Journal*, June 29, 1878, p.5.
[9] Ibid.

8 THE DEVIL'S RAILROAD

Construction of the Madeira-Mamore Railway had begun in early 1878, months prior to Ernest Morris's arrival in Brazil in mid-August. When he disembarked in Pará, there were already disturbing reports that the project was in trouble. In late February, the first laborers from Philadelphia had arrived at San Antonio on the Madeira river aboard the steamer *Mercedita.* San Antonio was projected to be the northern terminal of the railway, and was the place where construction would start. The *Mercedita*'s sea voyage had been long and very uncomfortable, as was its trip up the Amazon.

At San Antonio, the 220 workers found little more than the ruins of a town. They had to refurbish two abandoned structures to serve as their quarters; then they had to clear vegetation, build streets, and construct a wharf just to offload their remaining cargo. A decade earlier, a British attempt to survey Church's railroad had left behind a sawmill that the new American arrivals patched and put back into operation. Groups of men were assigned to survey upriver past the cachoeiras, but discovered the only way to move past the rapids was with native canoes and paddlers, who were hard to find. Construction material that the Americans planned to move past the falls had been packaged in large barrels and crates that did not fit the canoes, and so had to be broken down and repacked into smaller boxes.

Within a few weeks, the laborers found their provisions running short. They were unable to get fish from the flooded Madeira waters; and local game quickly disappeared into impenetrable forest. Their stores of flour, meal, and canned goods diminished at an alarming rate. Many of the men started to come down with malaria. A hospital building had not been raised yet, and their medical supplies were quickly diminished.

A second steamer, the *Metropolis,* had sailed from Philadelphia just two weeks behind the *Mercedita,* carrying an equal number of men, railroad materials, and supplies. The workers at San Antonio learned in late March that the *Metropolis* had wrecked off the shores near Currituck, North Carolina, on January 31, killing 102 men. The ships' owners and New York inspectors had overlooked the structural faults of the *Metropolis*; and the Collins' company overloaded the vessel. To compound errors, the captain mishandled the ship once it was in distress. Newspapers dwelt on recriminations for the disaster for months. The more immediate result was the loss of relief much needed by those already at San Antonio. With little choice, The Collins brothers hurried the launch voyage of a larger, third steamer, *The City of Richmond,* to bring much-needed aid to San Antonio.

At the same time, the Collins firm had another problem, based on the experience of the *Mercedita,* which had hit sand bars several times in reaching San Antonio. Help was needed to guide the supply steamers and schooners up the Amazon and along the reaches of the Madeira. What was needed were more adept vessels to tow the cargo ships through the narrow channels and shifting flood plains found along the route—in other words, tug boats. These craft could be found in the harbors of every America port, but none existed in Brazil. With their resources extended for the tasks already at hand, the Collins brothers decided to subcontract the tug service out to a different company, in exchange for concessions to exploit the resources of the upper Madeira river until the railroad was complete and the Bolivian national steam navigation company was ready to take over. The Philadelphia firm of Mackie & Scott, whose executives were associates of the Collins, offered to take the job.

The tugboats *Juno* and *Brazil* were dispatched to South America—a noteworthy, epic journey for two vessels not designed to navigate open seas. Meanwhile, Mackie gathered together a team of about a dozen geologists, physicians, mineralogists, chemists, and surveyors, engineers—and Morris, the naturalist—that were assigned to explore the upper Madeira and Mamore rivers and their tributaries to take stock of their commercial potential. With the railway not yet built, the plan was for three small steam launches to be carried in sections above the falls, where they would be reassembled; the exploring party would then work from the launches. It was projected that they would be in the field for one to two years.

Ernest Morris and the other professionals of the Mackie, Scott exploration team arrived in Pará in mid-August 1878. By that time, the third Collins steamer, *The City of Richmond,* and the two tugboats had arrived in Brazil, and had reached the Collins base at San Antonio. However, the new arrivals of laborers and supplies only slightly alleviated the problems there. Two hundred of the new arrivals were relatively unskilled Italian immigrants, who were assigned to make the rail bed that the track would be placed on. When they discovered their pay was much lower than the other workers, they went on strike. The strike was eventually forcefully put down by the handful of managers on the site, but not without leaving lingering tensions. Fresh food was still a problem, as was malaria and other illnesses. Some of the sickest men gave up the effort and made their way back to Pará, where Morris saw them wandering the streets. Since they had not been paid, many were left to beg to raise money to get home.

The bad state of affairs at San Antonio caused Mackie & Scott's exploration team to delay leaving Pará. The railroad project's misfortunes spread to Mackie & Scott's two tugs: with each tug's hull drafting nine feet, they proved as prone as the large cargo steamers to running aground on sandbars. More pertinent to Morris's role, reports from San Antonio cast great doubt on the prospect of transporting the sections of the steam launches over the falls—Morris himself expressed skepticism on this point in his private journal. In September, some of the exploration team members saw the writing on the wall and returned to the United States; Morris stuck it out until the end of October—mainly because he was battling another bout of malaria—but then resigned as soon as he felt well enough to travel.

Work stopped at San Antonio in late 1878, and the last American engineers left in early 1879. Of the 941 men who had gone to build the Madeira-Mamore railroad, 229 lost their lives, mostly to malaria. Seventy-five tried to walk out through the jungle and were never seen again. The firm of P & T Collins lost a reputed $800,000 and were forced into bankruptcy. The project was revived a generation later, in 1907; and completed in 1912, but with a much grimmer casualty count: perhaps 10,000 or more workers died to complete the railway. By that time, the rubber boom in South America was over (thanks to Henry Wickham's smuggled seedlings) and the rail traffic never brought the wealth that had been anticipated a half-century earlier. Today, Brazil, Bolivia and Peru are coordinating the damming of the Madeira, submerging much of the rusted track that was laid at such a dear price.

The collapse of the Mackie-Scott expedition was a disappointment for Morris—he not only wanted to explore the forests of eastern Bolivia, but also wanted to "cross the Andes, see far-famed Potosi, snow-covered

Illimani, and then I will be prepared with notes of interest enough to write about and talk of for years."[1] Morris received a very small stipend for all his trouble. He also asked for a good share of the trinkets and trade goods that the company had planned to use with Indians in Bolivia.[2] Then he took a steamer to Santarem, arriving on November 3, 1878.

His brother Frank was boarding on a plantation outside of Santarem with one of the Confederados, Reverend Richard Hennington, and had made no headway as an entrepreneur. Frank wrote to the *Indianapolis Sentinel*: "...for my part I would rather be hung in the States than die a natural death in South America."[3] The young Indianapolis law clerk, Jeremiah Johnson, had not even made it as far as Santarem. He deserted the Morrises just one week after arriving in Pará—it was too uncomfortable for his tastes.[4]

Ernest convinced Frank to move upriver with him to Juruti, his favorite little town. In Juruti they were greeted by Edward S. Rand, who had remained in Juruti ever since he and Morris moved there in January of 1878. Ernest Morris and Rand enjoyed meeting again, but Frank, despite Juruti's delightful setting, had become moody with what Morris called "the disease of discontent."[5] Frank had spent all of September and October collecting butterflies and beetles and had tired of that effort. Frank had written home that "...we [he and Ernest] would soon start up some river. I suppose I might as well go, as I can't get in a much worse place." [6]

Waiting out the rainy season for five more months in Frank's company did not seem appealing to Ernest; and though he was free from obligations to Mackie & Scott, he still had responsibilities to the *New York World* and to the orchid collector, Erastus Corning. Morris cherished a desire to ascend the Xingu; but after his Tapajós adventure he knew better than to attempt that during the upcoming seasonal flooding. The one advantage he did have was a perk he had received in Pará: a temporary free pass on the Amazon Steam Navigation Company's vessels. Frank showed little interest in further travel, so Ernest left his brother and Rand in Juruti, took a canoe to Óbidos, and then on November 16, 1878, boarded a steamer to Manaus.

Morris arrived in the city of Manaus on the 19th, after a dreary voyage full of bad food and obnoxious fellow passengers. After a day or two presenting introductions and making inquiries about good areas for orchid hunting, Morris started on a series of two or three-day canoe journeys up the Rio Negro. The best orchid areas could only be reached by boat; and he needed to get plants back to Manaus quickly to properly prepare them for shipping. These short expeditions provided material for his first column for the *New York World*; in it, Morris's love for the landscape is made plain:

> I remained in Manaus long enough to provide [i.e. secure] a canoe for my long voyage into the interior in search of orchids...the canoe was a large one, known in Brazil as an *igarité*, with a palm tolda or cabin, in which I stored my salt fish and farina. My crew consisted of two men and a boy of 15, and a villainous-looking set they were. After embarking bag and baggage and one of the men, who was half-drunk, we left Manaus with two paddling. The day was very hot, and the white sandy beaches were desolate. We passed no canoes and drifted slowly down the river without speaking a word; at the mouth we crossed and entered one of the smaller streams that intersect this whole region. As we slowly advanced I saw that the land through which we were passing was a flooded forest during the rainy season. The stream that wound its way through this wilderness was not deep, and every now and then we were forced to get out and find the channel. As I had by this time discovered that there was a black bottle in the canoe with too little rum in it and too much in the crew, a peremptory order to stop drinking was finally obeyed and the men resumed their paddles with more vigor. The forest bordering the stream was low, the trees covered with trailing dead grass, the debris of last year's flood. There were no houses on the bank, and no water to drink, for every stroke of the paddle stirred up the rank ooze of the bottom. As we advanced the stream widened and at the same time shoaled, and we were at last stuck fast in five inches of water only a mile from a lake of ample depth. The crew, except for the young Indian, were now hopelessly drunk, for they had got their week's pay before starting and had managed to get a supply of rum.
>
> Leaving the drunkards in the canoe and giving my boy the lantern, we waded ashore waist deep in mud, mindless of alligators [caiman], and set out on foot for a hut a half a mile away. The forest was very dark and the howling monkeys made it ring with their unearthly yells. In a few moments the storm was upon us. Great gusts of wind were accompanied by terrific thunder and lightning and followed by a deluge of rain. A tropical storm can not be exaggerated. The crashing of trees, the tremendous thunder, the deluge of rain, and the vivid lightning make an

impression from the sublimity of which the howling of the monkeys does not detract. Guided by the lightning (for our lantern had gone out), we trudged on, and in half an hour arrived at the low palm hut. Part of the roof was gone, but the place was still a shelter, and throwing ourselves on the dry sand in our wet clothes we listened to the distant peals of thunder until the pattering rain on the palm roof soothed us to sleep. On returning to the canoe the next day we found the storm had demolished the cabin, and that most of our provisions were spoiled; but we pushed on.

This one day's experience was repeated with variations for many weeks, during which I wandered this wilderness, often in want of things to eat and with rain nearly every day; but I succeeded in making a magnificent collection of orchids in the water paths of this great river. The canoe glides close to the bank and often the spreading branches of the trees interlace overhead, making night even of noonday. Countless species of sepoys (for by this name every woody climbing plant is known in Brazil) knit the whole forest into a tangled mass. Long aerial roots reach down from the tops of might trees; great *Bromelias* cover the branches and throw great spikes of flaming scarlet high into the air; orchids in profusion cling to every limb, and fern climb up the trees or wave graceful fronds from high in the air. Never have I (although accustomed to tropical scenery) been more impressed with the luxuriance and profusion of nature than here. Privations and hardships were worth undergoing for this grand sight. I have no language to describe it. At each turn of the creek there were always objects of beauty and wonder. Imagine a tree covered with plants of *Cattleya superba* in full bloom, the rosy purple petals and the deep purple lip, and in shading no two alike, the air full of perfume; the lovely little *Cattleya luteola* in full bloom, its plants covered with lemon-colored flowers. *Maxillaria* was not uncommon, hanging in masses many feet square. *Oncidiums* and *Brassavolas* were common. Of *Epidendrums* there was a great profusion—many in bloom, and a few quite as beautiful as the *Cattleya*. I hope I have given some idea, but it can only be a faint one, of the wonders that await the traveler who leaves the beaten road of Amazon travel and enters the unknown flooded forests of the Solimões and Rio Negro. It was with regret I left this naturalist's paradise, and with a feeling that I had done nothing. Years would not exhaust the flora of this region.[7]

Morris wrote this column after a series of canoe trips on the Rio Negro emanating from Manaus, from late November through December 1878. The Amazon splits into two separately named rivers a few miles east of Manaus: the northern tributary is the Rio Negro, where Manaus was situated, with its headwaters originating in the region between eastern Colombia and Venezuela. The Southern tributary is known as the Solimões, or Upper Amazon, and originates in southeastern Columbia, Peru, and Ecuador. In late December 1878, Morris took a steamer from Manaus that went east to the junction of the two tributaries and then proceeded westward up the Solimões, to the town of Tefé, Brazil, the largest settlement between Manaus and Peru. There he presented letters of introduction to Senhor Joaquim Batalha, a respected official and scholar who had once befriended Louis Agassiz more than a decade earlier.

L'intérieur du canot by Édouard Riou. From: Biard, François. *Deux anness au Brésil.* Paris: Librairie de L. Hachette, 1862. [Public Domain]

Batalha took Morris on an orchid hunting expedition up the Parana do Aranapu, a channel leading from the Japura river to the Solimões. As he had encountered on the Rio Negro, this area was all flooded forest lands, containing just a handful of houses built on stilts. At one of these houses, they stopped and met a trader who had come from up the Japura river—he had two children of the Miranha Indians that he was selling as slaves. The Miranha were a feared tribe on the Japura, with a bad reputation for warring against their neighboring tribes and selling captives—

or even their own children—into slavery. Slave trade among the native population was supposedly outlawed in Brazil, but the practice still flourished in the far reaches of the empire. Sr. Batalha considered the Miranha to be natives of Colombia, and therefore not subject to the edict outlawing the enslavement of Brazilian natives. The trader warned Morris against traveling up the Japura river; not only were the Miranha dangerous, but there were other hostile tribes to be found in the region. As proof, the trader gave Morris a native blow-gun and a pot of the poison pot curare, used by the Indians on the tips of their arrows.[8]

Submerged Forest by Franz Keller. From: Keller, Franz. *The Amazon and Madeira rivers; sketches and descriptions from the note-book of an explorer.* Philadelphia: J. B. Lippincott & Co., 1875. [Public Domain]

With a haul of orchid plants collected on this excursion, Morris and

Batalha returned to Tefé; and soon afterwards Morris boarded a steamer to return down the Solimões and Amazon with his cargo. He stopped at Manaus to change steamers, and while there witnessed a street procession that was part of a religious feast. As a statue of the Virgin Mary was paraded past, Morris doffed his hat, but refused to kneel as others in the crowd did. Seeing this, the priest in charge of the procession commanded Morris to kneel. In Portuguese, Morris explain that he was not Catholic. Infuriated, the padre ordered some nearby soldiers to arrest Morris. He was taken to jail, but when the chief of police arrived, he immediately released him. Morris communicated a complaint to the provincial president, and eventually had the satisfaction of hearing that the bishop of northern Brazil rebuked the priest in question.[9]

From Manaus, Morris made his way on a steamer to Óbidos, and from there took a large canoe back to Juruti, reuniting with his brother and Edward S. Rand. Morris had no specific plans at that point, but did have nearly 1000 orchids; 7000 butterflies and moths; and 14,000 beetles.[10] Rand, perhaps the most skilled orchid grower in the western hemisphere, assured Morris that all the plants would die if they were not properly cared for and transported back to the United States with personal attention. Frank had long-since tired of life in Brazil, and likewise urged Ernest to take him back to the United States.

Morris tarried in Juruti, relaxing in his favorite locale before facing the long journey north. Frank had done little specimen-collecting work while Ernest was in Manaus and Tefé, preferring to flirt with the native girls in Juruti. Ernest compiled a poem for Frank that played on the fabricated story from two years earlier, in which Ernest was said to have had a romance with an Indian Princess. Later, Ernest had copies printed up:

Francisco Morrises' Meeting Terresa, the Indian Princess

There was a youth who lived out West,
Francisco was his name,
He tarried first at Santarem,
To Juruti then came.

Spending the warm December days
Under soft sunny skies,
In roaming thro' the palmy woods
And catching butterflies

One day beside a crystal lake
He sat him down to rest,
Watching the brilliant dragonfly
Skim o'er its placid breast.

When suddenly before his gaze,
Beneath a palmy shade,
Light tripping o'er the ferny path,
Appeared an Indian maid.

In truth she was not over-dressed,
A single robe of white,
A little necklace, and long hair,
Black as the darkest night.

Amazed, confused, she stood awhile,
And did not dare to speak,
'Till he arose and softly pressed
A kiss upon her cheek.

She spoke in a soft liquid tongue,
He in a harsher tone,
But language is not from the lips,
Or framed in words alone.

For liquid glances of the eye,
Soft pressure of the hand,
A kiss, an arm around the waist
All nations understand.

Why tell the progress of their love,
Of meetings late at night,
When her papa was fast asleep,
And stars were shining bright.

So it went on from day to day,
'Till he uneasy grown—
Said, I must see more of Brazil,
Not Juruti alone.

So one bright day he sailed away,
Leaving his love forlorn;

'Twas always thus the way of man
Since first the world was born.

She wanders thro' the perfumed woods
Far lovelier in her grief,
And muses on that dream of love,
So sweet, yet ah! how brief.

At times upon the river bank
She weeps the long, long day,
And thinks of false Francisco,
Who wanders far away.[11]

Reluctantly, Ernest agreed to take Frank and their specimen collections back home to the United States, although he was already determined to return to Brazil within a few months. He now had a new ambition, in addition to ascending the Xingu—he wanted to explore the Japura River above Tefé and search for an unmapped channel connecting it to the Rio Negro.[12] The brothers sailed from Pará to New York, arriving April 28th, 1879. In contrast to the comic poem he had composed for Frank, on leaving Brazil, Ernest penned much more wistful lines, later published in the *New York World*:

The Bells of Juruti

High on the bluff, above the mighty river
Whose ceaseless current rushes to the sea,
Where from tall palms the sunbeams glance and quiver,
Perches the little village Juruti

'Neath smiling skies where gentle breezes blowing
Temper the tropic sun's too ardent ray,
And time glides on, in gentle ripples flowing,
His course one long, enchanting summer's day.

Afar the world with all its wild commotion,
Its waves of strife whose dashings never cease,
Afar the tossing of life's restless ocean,
Here 'neath the palms is quiet, rest and peace.

The little palm-thatched church uprears forever
Its simple cross to ever smiling skies,

And when the sunbeams slant across the river,
Far o'er the waves its sacred shadow lies.

Then chime the little bells, their music falling
In gentle cadence on the aegis shore
To voyagers on the mighty river calling
Rest, for the evening cometh; toil no more.

So in life's eve, tho' all the daylight toiling
To urge thy little bark 'gainst rushing sea
Which casts thee back, thy strongest efforts failing,
Listen for holier bells than Juruti.[13]

Just one week after Ernest T. Morris stepped off the *City of Rio de Janeiro,* the May 1879 issue of *Scribner's Monthly* rolled off the press, carrying the first installment of Herbert H. Smith's series on Brazil, an article entitled "The Metropolis of the Amazons," which was a twelve-page portrait of the city of Pará, illustrated by J. Wells Champney's subtle engravings. Morris might have envied the undeniable writing skill Smith exhibited, as well as Champney's sketches, but would have dismissed Pará as an uninteresting subject. However, in New York he encountered people who confused his writing for the *World* with Smith's for *Scribner's,* mistaking one young naturalist just back from Brazil with the other. When he arrived back in Indianapolis, he made a point of telling the Indianapolis Journal that he had received an offer from "an Eastern publishing firm" to write a book about his earlier journey up the Tapajós.[14] However, he was anxious to get back to the Amazon, and left that task for some future date—a decision he may have later regretted.

---

[1] "Ernest Morris," *Indianapolis Journal,* June 29, 1878, p.5.
[2] Brief mentions of Morris, as well as complete details of the mishaps of the 1878-1879 Madeira railway debacle, can be found in: Craig, Neville B. 1907. *Recollections of an ill-fated expedition to the headwaters of the Madeira River in Brazil.* Philadelphia: J.B. Lippincott.
[3] "Life in Brazil," *Indianapolis Sentinel,* December 17, 1878.
[4] "Personal," *Indianapolis People*, September 21, 1878, p.1.
[5] Morris used this phrase to describe Frank's mood in a journal entry.
[6] "Life in Brazil," *Indianapolis Sentinel,* December 17, 1878.
[7] "Up the Amazon River," *Indianapolis Sentinel,* March 1, 1879.
[8] "Trade in Children," *Marion Star,* April 4, 1879.
[9] "Ernest Morris Home Again," *New York World,* April 30, 1879, p.2.
[10] Ibid.
[11] One printed copy of this poem exists in the collection of Ernest T. Morris

papers held by descendant Marianne Lanman Salaymeh.
[12] Morris might have been able to accomplish this, if he had the opportunity. Although it is not a connecting channel, the headwaters of the Rio Uneiuxi are within a mile or two of the Japura, and descend to the Rio Negro.
[13] "The Empire of Don Pedro," *Indianapolis Sentinel*, May 8, 1879.
[1414] "Ernest Morris," *Indianapolis Journal*, May 19, 1879, p.3.

9 THE AMAZON POISON

Ernest Morris arrived back in Pará to start his fifth journey to the Amazon on July 3, 1879. His intention was to return to Tefé on the Solimões; and from there he would take a canoe and twenty Indian paddlers to ascend the Japura River. He would attempt to get past the Araracoara falls into a little-known area said to be rich in mineral wealth and cinchona, the plants from which quinine was derived. Unless Morris had very dated information, he probably was not interested in hunting for cinchona trees. More than a decade earlier, Clements Markham of Great Britain had recruited individuals—including naturalist Richard Spruce—to collect seeds and seedlings of the highest-quality cinchona trees in Bolivia, Peru, and Ecuador. By the 1860s, native bark collectors had depleted so many stands of cinchona that the world's supply of quinine was threatened. Markham's agents successfully brought seeds from the Andes that were transplanted to Britain's India plantations and the Dutch's Java holdings. By 1880, the countries of the Andes had already seen their cinchona exports plunge by more than a third—and none would have welcomed any further cinchona prospecting by Morris.

Morris thought that far up towards the headwaters of the Japura (known as the Caquetá in Spanish-speaking Colombia), there might be a connecting channel to the far reaches of the Rio Negro. He was mistaken in this guess, since those sources are separated by several hundred miles--though one tributary of the Rio Negro reaches within one or two miles of a section of the Japura at a much lower point in its course, still in Brazil.

However, before going back to Tefé, which lay about 1500 miles distant from Pará, Morris first took a steamer from Óbidos, and from there a canoe to Juruti. There, he had a happy reunion with his friend Edward S. Rand, who had been residing in the little village for over a year and a half. Rand had been envious of the profusion and rarity of orchid species that Morris had collected from the flooded forests near Manaus on his previous trip earlier in 1879. Though Juruti had its charms, the flora of the upper Amazon offered better opportunities for a collector than the lower or middle reaches of the basin. Rand convinced Morris to accompany him up the river to Manaus, where they could both easily make short trips into the nearby forests, especially the area between the Taruma-Minim and Taruma-Acu streams. Morris agreed, and they arrived in Manaus via steamer on August 4, 1879.

After settling in Manaus for a few days, Morris and Rand made a four-day orchid-hunting expedition to the Tarumas, a trip which Morris later described in a column printed in the New York World on September 30, 1879. Because Morris was headquartered in a major town on the Amazon, his letter had reached New York with unusual speed. Their plant collecting came at the height of the dry season, which was not the best time to find flowers in bloom. Morris ended his column for the *World* with an indication that he would be heading for the Japura river; but after posting that letter, he and Rand encountered a British ornithologist, John Hauxwell, who had been visiting the Amazon basin for decades, collecting bird specimens. Hauxwell had just returned from Pebas, Peru, which was situated on the Solimões (Upper Amazon) another 500 miles beyond Tefé, and sang the praises of the Pebas region for being full of unique flora and fauna.

Hauxwell succeeded in firing up enthusiasm for Pebas so well that Morris decided to change his plans and postpone going up the Japura, and instead decided to head into Peru to Pebas. Rand agreed to go with him. They departed Manaus on August 28 and took ten days to steam 1000 miles up the river to Pebas. All the Amazon steamers required enormous amounts of wood as fuel, so long stops were made every day. Often these delays took hours, as there were no loading docks, and so sticks of wood had to be hand loaded from the shore, one-by-one. On September 9, the steamer stopped at one such wood pile, surrounded only by dark forest and with no people in sight, and the captain declared "...and this is Pebas." Morris and Rand were dumbstruck--until the captain explained that the village itself was situated a quarter mile up a stream that flowed near the wood pile.

Rand and Morris found Pebas to be a miserable little village of 150 residents. The first surprise for Morris was that no one understood his Portuguese, for they were now in Peru, and the language was Spanish. There were shown into their lodgings: four walls made of vertical poles separated from each other by gaps of four inches, with a roof made of palm thatch. That evening they slung their hammocks, but were constantly woken by the sounds of barking dogs and grunting hogs, who freely roamed about the village. In due time, they moved to two rooms in a mud-walled house, said to be the prettiest house in Pebas.

For the next month, Rand and Morris spent their days collecting orchids and butterflies as best they could. Even though it was not the rainy season, nevertheless it poured every day, flooding with mud the few paths that led into the rainforest. Their existence became monotonous. On October 19, Morris wrote in his journal: "Ah! But I am tired of this stay in Pebas. O why did I not go to the Japura? I long for some excitement and have led a quiet life since I returned from the Tapajós—but will soon be in for it."

Interior of Primæval Forest on the Amazons by Josiah Wood Whymper. From: Bates, Henry Walter. *The Naturalist on the River Amazons.* London: John Murray, 1863. [Public Domain]

As they settled into their dull routine, Morris spent evenings working on his book manuscript describing his journey up the Tapajós three years earlier. In mid-November, he accepted an invitation to join the captain of the steamer that stopped at Pebas to go upstream another hundred miles to the town of Iquitos. Morris went there and stayed for three days, remarking only that the town was dirty and dilapidated, but that the young women were exceptionally beautiful—and loose with their morals. Iquitos was as far as the Brazilian steamers went; the steamers operated by Peru went further up the river, but Morris had no free pass on those.

After returning to Pebas, Morris continued work in his manuscript. By December 15, he had produced 186 handwritten pages; he wanted to finish the project before starting out on a search for the "excitement" he craved. Since arriving in Pebas he had spoken with many of the Indian residents, most of whom identified themselves as Yagua tribesmen, although they no longer lived in the traditional village. The Yagua were famed as blow-gun hunters, and had perfected the curare toxin used on their darts so that it killed swiftly without tainting the game. They traded terra cotta pots of their poison throughout the upper Amazon. Morris wanted to be taken to a Yagua village so that he could see and record the process they used to produce their curare, known in that region as *wourari*. Then, from the Yagua village, Morris intended to continue northwards through the forest until he came out at the Ica river, which he could descend to arrive back at the Solimões river. Morris had only a rough idea of the distances, but in fact, it was about twenty miles from Pebas through the forest to the Yagua village; and then another fifty or sixty miles to reach the Ica, through what is now the Zona Reservada Yaguas.

Fabricaion du poison by Édouard Riou. From: Biard, François. *Deux anness au Brésil*. Paris: Librairie de L. Hachette, 1862. [Public Domain]

Morris's intent to reach and descend the lower section of the Ica came to him at the same moment that another naturalist, the Frenchman Jules Crevaux, was making a thorough exploration of the Ica by steamer, ascending it to its westward reaches, where it formed the border between Peru and Colombia, where it was known as the Putumayo river. Morris had met Crevaux about a year earlier. Crevaux had shown Morris some of his collections, which did not impress Morris—he noted that he did not see many specimens that he had not collected himself. When the two met, Morris seemed unaware that Crevaux was planning a trip up both the Japura and the Ica—both of which Morris made his goals in 1879. It would be interesting to know if the two discussed the tributaries of the Upper Amazon and so might have influenced each other's plans.

While Morris waited in vain to find a Yagua Indian that was willing to guide him to a native village, on December 16, 1879, Edward Rand decided that he could stand no more of Pebas and boarded a steamer heading back down to the lower Amazon. Morris stayed behind at Pebas, and immediately missed the company of the older man (Rand was forty-five in 1879). However, within a few days he finally found a guide, who took him on paths through the forest twenty miles from Pebas to a Yagua maloca. There Morris stayed two days to observe the *wourari* being made. The Yagua used a complicated process involving a dozen or more different plants, but the active ingredient was found in one woody vine, *Chondrodendron tomentosum.* From the Yagua maloca, Morris was led to a different settlement, Ticuna, for a short visit; and then returned to Pebas after a journey of five days. At the villages, he found no one willing to show him how to get to the banks of the Ica; moreover, Morris was beginning to feel ill, so his grand Ica exploring venture never materialized.

On his return to Pebas, Morris stopped at the village of Cochiquinas to recuperate. Like Pebas, Cochiquinas was inhabited by about fifty detribalized Indians and *mestizos* (called *mestiços* in Portuguese). He described the village to his *New York World* readers:

> The *teniente*[1] of Cochiquinas was an old Indian who had a family of thirteen daughters, all old enough to be useful and who were all india-rubber gatherers. They wore many bracelets made of lizard skin on their arms, to give them, as they said, strength. The land to the south of the village is *terra preta*, or black land, and intersected by numerous cold, clear-watered streams. Here the trees seemed alive with monkeys and tigers [jaguars] were very numerous. In these dense forests I have wandered for days, accompanied only by a small boy, whose duty was to carry a gun

or an insect-net. The wonderful luxuriance which nature here displays in the trees and plants has often bewildered me, although accustomed to tropical scenery. So beautiful are the flowering vines and trees, so strange are the fruits, that one never knows where to begin...[2]

Village of Cochiquinas on the Right Bank of the Amazon by Édouard Riou. From: *Le Tour du monde: journal des voyages et des voyageurs.* Paris: Librairie de L. Hachette, 1866. [Public Domain]

Back in Pebas and low on funds, Morris found himself stranded. His free pass on the vessels of the Amazon Steam Navigation Company had expired. Stuck in Pebas for the month of January 1880, he spent his time catching butterflies, stuffing birds, and flirting with the girls of the village. In his journal, he alternately complained about the sameness of his days; and took pleasure in his unfettered life. On the 15th of the month he sent off 1500 butterflies to Erastus Corning, hoping that sooner or later he would receive payment.

On February 8th, the Peruvian steamer *Morona* stopped at Pebas, and its owner, Don Carlos Mourraille, offered Morris the opportunity to join the steam launch *Mairo* on an expedition to determine the reach of navigation on the Napo river, a tributary of the Amazon that emptied into the river between Pebas and Iquitos, but which originated hundreds of miles northwest, in Ecuador near the foothills of the Andes. Morris jumped at the opportunity and promised to meet the *Mairo* in Iquitos in late February. He expected he might catch a later steamer from Pebas to

Iquitos, but when it failed to materialize, he hired a canoe and made a five-day paddle upriver to Iquitos.

Upon arriving in Iquitos, Morris learned that the *Mairo* was not yet supplied for its trip up the Napo. So rather than idle away several weeks, Don Mourraille suggested that Morris should take passage on the Peruvian steamer *Pastaza* during its regular journey up the Ucayali, to the terminal point of navigation at Roaboya, some 500 miles from Iquitos. The *Pastaza* plied this route twice a month, carrying few passengers and very little cargo—it was subsidized by the government mainly for appearances: to exert Peru's sovereignty over the length of the Ucayali. While this offer was not quite a voyage of exploration, Morris was still interested—it would allow him to see another one of the great source tributaries of the Amazon; the Ucayali traversed the same area that he had once dreamed of while reading Mayne Reid.

Morris described his journey up the Ucayali in a column for the *New York World.* The most interesting items he noted were about the Indian tribes inhabiting the different sections of the Ucayali. For instance, shortly after entering the Ucayali, the steamer passed the Rio Tapiche, a wildly meandering river that was home to the Matsés tribe, known for making war of neighboring tribes on the Ucayali. The Matsés are also known as the Mayoruna, but Morris used the even earlier term for them that the naturalist Bates had used: the Majerónas. Days later, the steamer stopped at the abandoned mission of Sarayaen, which now was inhabitated by barely fifty Mashco-Piro people. Today, contact with this tribe is proscribed by law in Peru, but Morris wandered the village and obtained a list of fifty words in the dialects of many local tribes: the Conibos, Shipibos, Mashco-Piros, and Asháninka. What Morris later did with that list is unknown, an unfortunate loss.

After five days, they reached Roaboya, the terminus of the steamer route. Morris noted that in rainy season, the river could be navigated much further south, to the point where the Rio Pachitea entered the Ucayali. Morris mentioned that only Jesuits ventured further up the Ucayali, and at their peril. Fifteen years earlier, two Peruvian officers had been killed in a battle with Cashibo warriors; and were rumored to have been cannibalized. The return trip to Iquitos took another five days, the *Pastaza* anchoring there on March 9, 1880. Three days later, Morris boarded the small steam launch *Mairo,* which was now prepared for its journey of exploration up the Napo.

The *Mairo*'s Napo expedition was the subject of another column that

Morris wrote for the *New York World,* offering Morris's signature blend of derisive views of civilized and uncivilized humanity; gripes over personal discomforts; and awe of the natural world. Because this article is a prime example of Morris's *New York World* columns, it merits reprinting here:

Up and Down the Napo

On board the *Mairo,* Rio Napo, Peru, April 20 [1880]–Twenty-five hundred miles above the mouth of the Amazon, and sixty miles below the village of Iquitos, Peru, is the mouth of an almost unknown river, the Napo. It is to Prof. James Orton, who descended this river from Ecuador some thirteen years ago, that we are indebted for the only reliable information yet made public respecting the so-called "Napo country" and Rio Napo. The first steamer that ever entered this river was the *Mairo,* sent by the Peruvian Government for the purpose of exploration, but for some unknown cause she returned in a few days to Iquitos. I have examined the journal from this expedition, but obtained little information from it.

In a previous letter to the *World* I have endeavored to explain how utterly ignorant of their own country are the inhabitants of this region. Geography is unknown in eastern Peru, and I never yet have been able to make my acquaintances here clearly understand why I should not know their relatives who reside "somewhere in Nueva York."

For several months past it was known in Iquitos that the "Peruvian Company" would again send the *Mairo* (which, by the way, is an American-built boat and was brought to Peru by the somewhat well-known Admiral Tucker) to the Napo, solely for the purpose of ascertaining how far the river was navigable for steamers. On March 17, the usual crowd of idlers gathered on the bank to watch our departure, the anchor was soon raised, and amid the shouts of "the happiest people of Peru" we rapidly steamed down the Amazon. We had a curious crew–Peruvian captain, English engineer, Chinese cook and seventeen deck hands of all ages, sizes, and colors. The *Mairo* is a small side-wheel boat, and when loaded draws but three feet of water. From stem to stern, about twelve feet above the deck, the boat was roofed with thin pine boards on which many of the crew spread their blankets and slept in the open air. Others used the coal bunkers, while our Chinese cook slept on the dinner table. As we had left Iquitos late in the day, we stopped for the night at a wood station near the mouth of a small river called the Nanay. We were kindly invited to a home by the owner, a woman and a perfect Amazon, who in ten minutes smoked all my tobacco, knew where I came from and where I was going, whether I was married or single, and the rest of my private history. She invited us to bring our beds up to the house, saying there were no mosquitoes, but we preferred to sleep on board and be eaten by the mosquitoes rather than be talked to death by a woman. During the night we embarked 2,000 sticks of wood, and on the following morning were again steaming down the broad river.

At 10 a.m. we entered the upper mouth of the Napo. This river enters the Amazon from the north by two channels, which are separated by a long, low island, subject to overflow. Half a mile above the mouth is a collection of four houses, called Destacamento. Here we anchored and sent our small canoe ashore for a pilot. The canoe soon returned, bringing an old Indian, who smilingly shook hands with all on board and said he knew the river and could pilot us to the mouth of a tributary stream, the Aguarico, about six weeks' journey in a canoe. A bargain was made with him immediately, and the captain joyfully went to take his afternoon nap. But a few moments after embarking the old pilot, and while steaming along rapidly, there was a sudden rattling of dishes; the boat lurched to one side, and we grounded with a shock in two and a half feet of water, amid the *carambas* of all on board. The pilot seemed very much astonished, and said he had no idea that there were any bars in this part of the river. When asked when he made his last trip to the Aguarico, the old Indian replied, "Forty-eight years ago." "What? *Caramba*!" Nothing more was said, and after some hard work we again backed into deep water; but thereafter we kept the sounding line going, and the old pilot was never seen to smile again, for we kept him busy from morning till night shelling rice, cutting wood, and washing table linen.

Twelve miles above the mouth we arrived at the small settlement of Magua [Mangua], on the eastern bank, twenty-five feet above the water. It is a charming little place of seven well-made houses inhabited by civilized Indians (I use the word civilized in an Amazonian sense–i.e. those who wear trousers and occasionally a shirt). Surrounding the houses were numerous Punpunha palms,[3] with ripe bunches of red fruit hanging among the crown of dark-green leaves. This fruit is eaten boiled and is a good substitute for the potato. The houses were neat and clean. The men were hardy-looking fellows, but the women, I regret to say, were positively ugly, with hands and faces smeared with black dye. The dress of the women consisted of a narrow piece of dyed cloth worn around the waist, but what they lack in cloth they make up in beads. The language spoken is called Inca[4], and indeed all the Indian population of eastern Peru speak Inca, which corresponds to the língua geral of Brazil. During the short time we stopped at Magua I caught twenty-five different species of butterflies, and I believe it would be a good place for a naturalist. The forests which hem the little village are very rich in tropical plants. Of orchids, I saw many varieties, principally *Oncidiums*. It may be of interest to the growers of these beautiful plants to know that one reason why so many of the Amazonian orchids do not survive in American hothouses is that they are subjected to entirely too much heat. So cool is it at night on the Napo that we wear a thick coat and sleep under a heavy blanket.

In thirteen hours we arrived at Mazán, a ruinous settlement of four houses on the western bank. Many of the inhabitants were suffering from chills and one man was dying from the bite of a poisonous snake. The poor fellow begged piteously for a *remedio*, but we had none to give him. A few miles above this settlement we pass the mouth of the small river Mazán, which enters from the west. Its banks are inhabited by the wild Iquitos Indians.[5] All day we have been winding in and out of narrow channels, or steaming close to the bank, where we often see brilliant birds or flocks of monkeys. At night we pass three houses known as Cayapas,[6] and on the following day, March 20, thirty-one hours of steaming time from Iquitos, we reach Tuta Pisco[7] (nightbird), the last settlement on the lower Napo. Above Tuta Pisco for a distance of over four hundred miles there are no settlements and the banks of the Napo are inhabited only by savages. On a point of land, composed of red clay, which juts far into the river, are several roofed stockades inhabited by 120 Indians of at least five different tribes. These Indians are employed by Señor Harra, an Ecuadorian, and owner of this settlement. We entered his house, which was no better or cleaner than those of his dependents, though he is said to be a wealthy man, and is engaged in collecting gold dust and rubber. He was very much opposed to opening the river to navigation, and expressed his opinion in a most surly manner. Hs wife, however, was very glad to see us, and asked whether we would take a cocktail (spoken in plain English) or a punch. On being told the latter she gave us two dozen eggs to beat while she procured cachaça [8]rum and sugar. Filling several large bowls with a hot mixture—I suppose it was punch, an Ecuadorian punch—she passed them around saying "*Sabi*" ("your health")[9]. Steaming hot punch was anything but refreshing with the thermometer ay 98 degrees in the shade, but we partook of her hospitality.

The forests surrounding Tuta Pisco are full of *marfim vegetal*[10] or ivory palms, which yield the vegetable ivory of commerce. This palm attains the height of only twenty feet with a crown of stiff dark leaves, which are used by the Indians in thatching their houses. The fruit grows in clusters, enclosed in a rough capsule. A tree will yield from fort to sixty nuts. On the lower Napo, at Magua and Mazán, these nuts are collected by the Indians, who sell them to Señor Harra for four *reals* (forty cents) per barrel. This enterprising person disposes of them in Iquitos for $4 per barrel. Indians are often compelled to take goods in trade at exorbitant prices, and it is to the patron's interest to keep them always in debt. An Indian once in debt virtually belongs to his patron. There is no country where there are so many unprincipled traders as on the Amazon and its tributaries.

We remained twenty hours at Tuta Pisco. While here I obtained two skulls of wild Indians known as Tuta Piscos.[11] During the evening, while busily engaged in cleaning my Indian skulls, we heard the low tap of a drum in one of the adjoining houses. A dance was going on. Calling our pilot to act as interpreter, we entered the house, which had but one long, dimly-lit room. On either side, seated on the ground, were some fifty persons of both sexes daubed with black dye. The music and dance ceased on our entrance and I plainly saw they did not like our intrusion. Nevertheless, a few wished us *boa noite* and the women offered us large wooden bowls of *masato,* which is made by the women, who chew the yucca and throw it into an earthen pot, where it is allowed to ferment. When fresh it is not intoxicating, but the Indians prefer it when it is about a week old, and it is difficult to tell at the end of that time what the *chicha*[12] jar contains—*masato* or Limburger cheese. Should one look into the jar he will see a moving mass, or as Prof. Orton expresses it, "writhing articulates."

So much *chicha* is used in the little villages of Cholquinas and Maucallacta, of which I have written the *World*, that there is not a good set of teeth among the women. We gave offense by not drinking the *masato*, and the dancing was not resumed during our stay. No sooner had we left the house, however, then we again heard the tap of the drum, and as we had determined to see that dance we stole through the woods to the rear of the house, where we witnessed their movements unobserved. Six couples danced at a time, the men all facing their partners and keeping step to the monotonous tap of the drum. Not a word was spoken by the dancers and not a smile was to be seen on their faces. Returning to my narrow quarters on board the boat I found many of our crew stretched across the wood sleeping off the effects of *chicha.*

We did not leave Tuta Pisco until late the following morning, when Señor Harra gave us an Indian to act as pilot. He came on board in his paint and with a jar of *masato* under his arm, which we quickly dropped overboard. Bidding adieu to our friends, after politely refusing another "Ecuadorian punch," we steamed slowly up the western bank. The more I see of the Napo the more I am impressed with it. Its picturesque islands are far more beautiful than those of the Amazon, and there is a greater diversity of tropical plants. Seen from a distance these islands present but a wall of living green, yet all the apparent sameness vanishes when the steamer nears the bank, and each moment of progress reveals wonderful variety. The vines which climb up the bank or trail in the water are often a mass of morning-glories, or are covered with a multitude of red or white flowers. Above the flowering vines are the palms, shaded by the tops of great trees, many belonging to the great family of *Leguminosae.* On a high point of land we saw many of these great trees in bloom, their wide-spreading tops one mass of colored flowers.

On the evening of March 22, we anchored in front of a high bank called Puca Urcu. Villavicencio[13], an Ecuadorian, in a work published in 1850 [1858], states that steamers can here be supplied with coal. So we anchored early in the morning, purposing to explore the surrounding country. Early the following morning the captain, with ten bushwhackers, armed to the teeth, entered the forest, while the pilots and I lowered a boat and paddled along the high bank. About fifty yards below our anchorage we found a narrow creek. The land bordering this creek, whose waters were black, was about twenty-five feet high, and here we saw a strata of clay entirely new to us, being pure white, free from sand and exceedingly plastic. At Iquitos and Pebas one sees a thin strata of lignite, which the inhabitants of the latter village have told me was carbon nueva or new coal[14], but we at Puca Urcu saw no signs of lignite or bituminous shale, but in all probability it exists and has been mistaken by the Indians as real coal. Leaving Puca Urcu we steamed slowly all morning. The river, as usual, is filled with islands and bars. At 11 a.m. the thermometer was at 82 degrees, but the hottest part of the day was between 2 and 3 p.m., when the thermometer often registers 96 degrees. This afternoon (March 23) we had a heavy thunderstorm, accompanied by wind and rain, which was quite refreshing and cleared the boat of pium, a small fly whose bite draws the blood. Our hands and faces are black with their punctures. At 5 p.m. we reached the mouth of the river Chia Curaray[15] (chia signifying little) which enters the Napo from the west. Some forty miles from Chia Curaray we entered the Napo's largest tributary, the Curaray Grande,[16] eighty four hours' steaming time from Destacamento, or the mouth of the Napo. This tributary, which also enters from the west, is 200 yards wide, with seven fathoms of water and a sluggish current. Its banks are higher than those of the Napo and slope gently to the water. We explored the river for a considerable distance and have no doubt that a small steamer could ascend it for twenty days. During the dry season the Indians of Señor Harra of Tuta Pisco wash gold dust from the sands. There are no settlements on this river; and its banks are inhabited by savages.

Soon after leaving the mouth of the Curaray we grounded in midriver. After hard pushing and pulling we got off the bar and steamed to the bank, where we spent six hours in cutting wood. This wood-cutting delays us very much, though it gives us the opportunity to hunt. I never saw game more plentiful than in the forests of Napo. Monkeys, wild hogs, and deer abound, and there are also numerous game birds. While cutting wood we entered the forest, and had not gone 200 yards from the bank when we heard the call of the maquisapa monkey.[17] The Indians imitated his cry, and soon we saw them slowly approaching us, swinging from branch to branch. Monkey-shooting is great sport, but when one who has only wounded a large monkey watches it picking out the pellets of shot and listens to its plaintive whining, he feels that he is anything but a true sportsman. The maquisapa, or long arm, is the largest monkey in eastern Peru. The color is black on the back and gray on the belly and under the arms and legs. The hair is long and coarse. We had monkey curried, boiled, and fried for dinner. Monkey meat is good, but we can never get over the idea that one may possibly be eating his ancestors.

We steamed until late in the night. It is a pleasure to lie full length on the top of the boat and listen to the howling imps of Satan, the guariba monkeys, and watch the fire-flies and bats. Occasionally we hear a grating noise, which we know is made by a tiger [jaguar]. To me these night voices add a charm to these forests. The river looks most beautiful and there is something more than mere poetry in "a starry night in the tropics." One never feels the want of words and his own insignificance so much as when he attempts to describe the beauties of nature.

After forty-seven hours of steaming time above the Curaray Grande, we pass on the right hand or eastern bank the mouth of the river Santa Maria. It does not deserve the name of a river. In Brazil this stream would be an igarapé (creek). The Santa Maria takes its name from a tribe of Indians who wander along its banks.[18] Since leaving the Curaray, we have found no less than six fathoms of water, which is now becoming muddy.

On April 2, thirty-two hours of steaming time above the Santa Maria, we entered the river Aguarico (rich water) which enters the Napo from the east. This river rises in the Cordilleras of Pimampiro[19] and flows directly south nearly 300 miles. In the rainy season it furnishes a vast amount of water to the Napo. It is 400 feet wide at the mouth, with eight fathoms of water. This river, as well as its largest tributaries, the San Miguel, Azuela, and Cofanes[20]--abounds in gold dust, specimens or rather ounces of which I saw in the possession of Señor de Sousa, whom I met in Iquitos and who lives in the Indian village of San Miguel on the Aguarico, thirty days in canoe above its mouth. To Señor de Sousa I am indebted for much information respecting this river. Not only does the Aguarico receive the three above-mentioned tributaries but numerous smaller streams, all of which he tells us are auriferous. The village of San Miguel is only inhabited by Indians of the Aguarico tribe.[21] From San Miguel there is a path leading to the river Ica. About a half-mile inside the mouth we cleared a large space of the underbrush, and on one tree nailed a large board bearing our names and the date of the expedition. I left one of the World's colored envelopes for the savages. The river Napo above the Aguarico bends to the north, and is at least a half a mile wide but shallow, and islands are not so numerous. At 8 p.m. we saw the campfires of the Indians, which were quickly extinguished on catching sight of the side lights of the steamer.

On Saturday, April 3, we espied a canoe moored near the mouth of a small creek. We anchored, lowered a boat, and, accompanied by the pilot, went in search of the owner of the canoe. Twenty yards up the creek we saw a rude shelter of palm leaves, and on our approach all the inhabitants fled into the woods; but the pilot, who up to this time had proved worthless, shouted in their language that we were friends. Two Indians, armed with long lances, soon made their appearance. They proved to be man and woman of the Santa Maria tribe. Both were dressed in short, sleeveless bark shirts reaching to the knees. The man accompanied us in his canoe to the steamer. He was terribly frightened. He said his tribe was encamped near the mouth of the Aguarico, and had watched the steamer from the woods. This was the first human being—though he looked more like a wild animal—that we had seen in thirteen days. He was below the average stature, with coarse long hair which reached below his waist. His face was round, with small obliquely set eyes, and his legs and arms were covered with white blotches. Through the pilot I obtained a list of words, and I am almost positive that the language spoken by the Santa Marias is the same as that of the Cotos, who inhabit the land near the mouth of the Napo.

Eleven hours above the mouth of the Aguarico we reached the river Yasuni, which enters from the west, with four fathoms of water at the mouth and a very strong current. On this river live the Shuripunos and Yasuni tribes of Indians. The former are said to have fishhooks made of gold. These tribes are but branches of the Zaparo nation,[22] which occupies all the land on the western bank of the Napo, between this river and the Pastassa[23]. This afternoon we fell in with a canoe of Zaparo Indians, who told us that there were some 200 of their tribe camped in front of the Jupitini[24], nine hours above the Yasuni. On April 4, we arrived at the mouth of the Jupitini, where on a long sandbar where the rude shelters of the Zaparos. On landing, we were met by the men and women; the children had fled into the woods. We were very much surprised at finding these Indians were a fine-looking, light-complexioned race of people. Many of the men have beards and were tall and muscular, with oval face, deeply set eyes and wearing a dull, restless expression. They showed no curiosity, but seemed anxious to trade with us. Some wore trousers, but the majority were dressed in long bark shirts. The women were smaller than the men and of a darker color. They evinced a good deal of curiosity (as might be expected) and wanted to handle everything they saw. They wore the usual strip of cloth and beads, and their fingers were covered with rings. Many also wore earrings, and all above ten years of age had a young Zaparo strapped to her back. I also saw many very old people. Their huts were of palm leaves, open on all sides. A thin strip of bark is used as a bed. The weapons are long lances and blow-guns with poisoned arrows. Here we bought animals and birds, hammocks and chambiri cord, all for a few beads, and our boat now is a floating menagerie. These Indians trade with the whites of the villages of Coca and Santa Rosa, a week's journey above in a canoe, and frequently make long voyages to the salt mines on the river Huallaga, the round trip occupying some five months. The curaca, or chief, at Jupitini tells us that the river Jupintini is navigable for thirty days in a large canoe, and that gold is abundant. The Shuripunos Indians, he says, have fishhooks and instruments made of gold. This possibly may be an El Dorado story.

Twelve hours above Jupitini we arrived at a new settlement of the Zaparos, called Sinchi Cheita[25] (strong party). Here the Indians had cleared about ten acres of land and had planted corn, cane, yucca, and bananas. The chief came on board accompanied by at least thirty persons. They were prepared to trade with us for poison. From this point northward, the Napo is about 300 yards wide and very shallow. On the 7th of April, we saw the high land of Panacocha Urcu, and were then only ten hours from the village of Coca, which is nearly 500 miles above the mouth of the Napo. Our steam pumps constantly filled up with sand, and at 2 p.m. we were hard aground with a sand bar forming around us. For five days we did not make ten miles. As fast as we got off one bar we got on another. I sounded the river near the outlet of the lake Yuturi, which is only half a mile above where we are aground, and found only three feet of water in the deepest part. Many of the crew have waded the river from bank to bank. We spent five days in trying to pass these bars, and by our watermark find the river to be falling.

We began our return voyage at 6 p.m. on the 13th, but did

> not reach Jupitini until the 15th, having grounded no less than ten times. After leaving Jupitini we steamed rapidly and, aided by a strong current, soon passed the mouths of the river Aguarico, Santa Maria, and Curaray Grande, and on the 19th reached Tuta Pisco, where we gave the boat a cleaning, and the men drank their fill of chichi. We also stopped at Mazán and Magua. Here the captain, through his carelessness, lost his diary overboard, and is now hard at work trying to copy from my numerous notes. If a future traveler should wish to see the Mairo's log-book, he will be shown but two pages of foolscap which reads: "Entered the Napo; cut 600 sticks of wood and grounded," and will obtain no other information. On the 21st of April, we again caught sight of the Amazon, and our voyage on the Napo was over. We shall never forget the lonely river with its picturesque islands, its long sandbars and miles upon miles of magnificent forests. To Don Carlos Mourraille, of Iquitos, I am indebted for much kindness, and I here wish to express my obligation.

Morris was offered a ride on the regular steamer back to Pebas, but also told Mourraille that he was interested in taking a steamer up the Huallaga river to the town of Yurimaguas. From there, Morris hoped to travel to the prime orchid-hunting regions of Peru, in the Alto Mayo forests between Moyobamba and Chachapoyas. From that point, he could go south to Lima to catch a ship home. Mourraille promised Morris he would let him know if that opportunity surfaced.

Morris spent the three months of May, June, and July back in Pebas, a village that he never grew to like, except for its wealth of flora and fauna. In early August, he received a message from Don Mourraille to come to Iquitos to catch a steamer up the Huallaga; but when Morris arrived in Iquitos on August 10, 1880, he discovered the situation had changed, due to the upheavals caused by the War of the Pacific. Peru's ally Bolivia had withdrawn from the conflict, leaving Peru in the position of battling against the forces of Chile on its own. In June, Chile had captured two Peruvian provinces and was laying siege to several ports, and threatening Lima. The Peruvian politician, Piérola, had successfully staged a coup-d'état in December 1879, but was badly mishandling the Army and the conduct of the war.

Mourraille told Morris that he did not know when the steamship company would be able to send a vessel down the Huallaga; and he strongly advised Morris against trying to reach the Moyobamba area. Others in Iquitos made the same cautions to Morris. After deliberating, he reluctantly gave up this goal, and decided to go back down the Amazon to Brazil—though he was not sure where. He had been suffering from fevers since July 1880, and was taking quinine regularly. If he could not go to the best orchid area in Peru, he figured he might as well return to the best orchid grounds in Brazil: the flooded forests on the banks of the lower Rio Negro. And so, Morris made the long steamer journey from Iquitos to Manaus, arriving at the end of August 1880. There he bought what few supplies he could, and in September boarded one of the steamers heading up the Rio Negro. He asked to be let off at one of the steamer line's wood stations, a half-abandoned little village called Tauapassu[26], located about sixty miles above Manaus, near the great Anavilhanas Archipelago (now a national park).

Morris spent a month in Tauapassu recuperating and collecting orchids, then was ready for another adventure. He now made it his goal to take a steamer to the terminal of navigation on the Rio Negro—Santa Isabel—and then hire a canoe to paddle twenty-five days further up the river, to the frontier region where Brazil, Colombia, and Venezuela meet along the Rio Içana. He boarded the next steamer upriver. As it passed the village of Moura[27], Morris saw that it was being protected by Brazilian soldiers, due to frequent attacks by Waimiri-Atroari Indians. Morris also noted that this section of the Rio Negro was very sparsely-populated, due to a prevalence of tropical diseases. He attributed the unhealthy conditions to the poor diet of the people, as these diseases had not yet been identified as insect-borne.

At the village of Tomar, Morris noticed many canoes tied up, and he approached the owners to see if they would be going further up river, above the many falls that blocked steamer navigation. By pure chance, he found the owner of one large canoe to be a young Spanish-born merchant named Germano Garrido y Otero, who was establishing himself as a virtual potentate far up the Rio Negro on its tributary, the Rio Içana. Garrido employed a crew of native Indian paddlers, led by a grizzled old pilot. This was everything that Morris could have hoped for—a repeat of his epic adventure navigating the cachoeiras of the upper Tapajós—now repeated on the cascades of the upper Rio Negro.

The crew of ten employed both paddle and a square sail, and had coils of palm fiber rope aboard to pull the canoe past rapids. The first rough water they encountered was at Santa Isabel, but with a good wind behind them, they sailed the canoe past the stretch with little resistance. However, soon the wind died down and the current was too strong to paddle against, so they had to haul out the ropes. Some of the men would paddle ahead in a small canoe, fasten the rope to a tree, and then pull the larger, heavily-laden canoe forward. Occasionally they encountered rubber traders; but the quality of the rubber trees on the Rio Negro was not nearly as good as those on the Tapajós. Still, Morris concluded that the rubber trade was responsible for depopulation of the upper Rio Negro, since it had ruined agriculture through neglect.

La cascade du Rio Negro by Édouard Riou. From: Biard, François. *Deux anness au Brésil.* Paris: Librairie de L. Hachette, 1862. [Public Domain]

When the small canoe was not needed to advance with rope, Morris commandeered it to explore the forests near the banks of the river; he found orchids, but often they were high out of reach, and he did not have time to retrieve them. He did sometimes bring back game, however:

> The woods along the bank were alive with toucans and parrots. I took my gun, embarked in a small canoe which serves as a tender and paddled to shore, where in less than ten minutes I killed enough toucans for our breakfast. The flesh of the toucan–a bird with a wonderful beak and beautiful plumage–is very good eating.[28]

Passing the hill of Jacamim, on October 21, 1880, they neared the village of Castanheiro, where Morris left letters that would later be taken downstream. They found the village to be mostly deserted, but Morris found an old man who regaled him with anecdotes about the naturalists

Spruce and Wallace, who had traveled this far up the river many years earlier. He also spoke of a guardian spirit of the forest, Aüeauló, "a woman with long black hair, armed with blow-gun and arrows, and who wanders the banks of this river, and until the Aüeauló is seen on some other river, the exceedingly rich forests of rubber and sarsaparilla will remain untouched."[29]

Adventure with Curl-Crested Toucans by Josiah Wood Whymper. From: Bates, Henry Walter. *The Naturalist on the River Amazons.* London:

John Murray, 1863. [Public Domain]

While camping at the foot of the Camanaús rapids, members of the canoe crew captured a *jacaré-tinga*, or white caiman (*Caiman crocodilus*). As was the case with piranha, Morris's writings gave only brief attention to this much-feared denizen of Amazon waters. On this occasion, the reptiles had more cause to fear man than they him:

> Over one of the camp-fires the crew are roasting with boisterous merriment a live alligator (jacaré-tinga), about five feet long. When I asked why they did not kill the animal before roasting, the pilot, who is always the spokesman of the party, answered that it would spoil the meat. The white alligator is highly relished by both whites and Indians. It differs entirely from the *jacaré-açu*, or large alligator,[30] rarely attaining five feet in length, and is distinguished from the larger species by its pointed nozzle, somewhat rounded tail, whiter color, and its freedom from the *acatinga* (smell). Though it is found throughout the whole course of the Amazon, it abounds more in clear-watered rivers and creeks. I have often found this alligator in streams of the high hills, miles away from any river or lake, and have frequently seen the skulls and bones in the forest. That it travels far and well on land there can be no doubt; and the Indians say that its eggs are deposited in the forests. The flesh resembles veal in appearance, but has a fishy taste. I left the crew to enjoy their feast.[31]

Catching an Alligator with Lasso by Franz Keller. From: Keller, Franz. *The Amazon and Madeira rivers; sketches and descriptions from the note-*

book of an explorer. **Philadelphia: J. B. Lippincott & Co., 1875. [Public Domain]**

Morris did not consider them much of a threat. "In some of the small, muddy, sluggish *igarapés* they are often very numerous. They seldom, however, attack man, but cause much loss to cattle owners, frequently biting off the legs of horses and cattle as they stand in the water, or large pieces from the neck as they drink."[32]

With great effort, the crew pulled the canoe past the worst two rapids of the Rio Negro: Camanaús and Sáo Gabriel da Cachoeira. Above that, they passed a half dozen more cachoeiras, but none contained the dangerous *furnas* or whirlpools found at Sáo Gabriel. On November 8, 1880, they reached the Rio Içana, and on the 11th arrived at their destination, the village of Sáo Felipe, where Garrido y Otero resided. Morris, along with several of the canoe crew, had experienced chills during the ascent of the Rio Negro, but at Sáo Felipe his condition turned into a full-blown fever that immobilized him. He lost one shoe of his only pair, and had to go barefoot. At one point while reaching for an orchid, his arms and legs were immersed in a toxic river sediment that was locally named *cauchy*.[33] It caused rashes on his extremities.

Morris spent the month between mid-November and mid-December making "a magnificent collection of orchids," but wrote little else of his time on the Içana in his journals. He had written several columns for the *New York World* detailing the ascent of the Rio Negro, but nothing about the Içana or his return down the river. The next time he is heard from is in letters written in February 1881, a month after he returned to the little village Tauapassu on the lower Rio Negro to recuperate. He wrote that he was suffering from severe jaundice, which could have been caused by any one of several tropical diseases that affect the liver. He bemoaned to a friend in Indianapolis: "I am growing old. The [past] two years which I hope to complete have not passed lightly over my head, yet I am still as enthusiastic as ever, and still am looking forward when I will be able to continue my studies in another part of the world."[34] He also mentions that while recovering his health in Tauapassu, he was finishing his manuscript about his trip on the Tapajós, and hoped to have it published in the fall of 1881. In March, he left Tauapassu and journeyed all the way down the Amazon to Pará, bringing with him 1350 plants. He had 900 of them loaded on the British steamer *Glenlogan,* bound for New York; and left 450 plants in the care of his friend Edward S. Rand, who had settled in Pará. Ernest T. Morris and his collection arrived in New York on April 23, 1881.

[1] Literally, a lieutenant; in this case, a local man appointed by the region's governor to manage village affairs.
[2] Morris, Ernest. "Ernest Morris in Peru," *New York World,* January 27, 1880, p.5.
[3] Morris is referring to the Brazilian common name of *Bactris gasipaes*: pupunha
[4] Now commonly known as Quechua, the most widely-spoken Indian language of the Americas.
[5] It is unclear what specific people Morris is referring to. The generic Iquitos name was given to detribalized Indians living in settlements along the Solimões (Upper Amazon) near Iquitos. The enslavement of native peoples by the rubber industry so disrupted tribal coherence that even by 1880, Morris was likely encountering remnant populations of various peoples.
[6] Cayapas is not on any current maps of the lower Napo.
[7] Now written as one word, Tutapisco.
[8] Cachaça, a Brazilian rum
[9] It is not clear whether Morris misheard the toast word or mistranslated it.
[10] A common name for *Phytelephas*
[11] It is not known what recorded tribal group Morris is referring to as Tuta Piscos.
[12] A fermented drink made from maize or yucca, often after being mixed with saliva to break down the starches.
[13] Dr. Manuel Villavicencio (1804-1871) was not an Ecuadorian, but was a Peruvian physician and naturalist.
[14] Lignite is also known as brown coal, a rock formed from condensed peat
[15] Morris means the Rio Tacsha Curaray, also known as Rio Chico Curaray, i.e. "Little Curaray".
[16] On most maps called simply Rio Curaray
[17] Spider monkey, *Ateles belzebuth*
[18] Today this tribe is known as the Secoya.
[19] The Cordilleras are mountain ranges in the canton of Pimampiro, northern Ecuador.
[20] While the Azuela and Cofanes are tributaries of the Napo, they are no more significant than a dozen other named tributaries that flow into the Upper Napo. The Rio San Miguel, now called the Rio Coco, is a major tributary of the Napo.
[21] By Aguarico tribe, Morris likely means the Cofán people, in what is now the Cuyabeno Wildlife Reserve.
[22] Known today as the Sápara people.
[23] Spelled now as Pastaza.
[24] Morris is referring to Rio Tiputini.
[25] Spelled today as Sinchi Chicta
[26] Tauapassu is not on any current maps; it may be in the same location as the current small settlement named Tauatu.
[27] Sometimes spelled as Moreira.
[28] Morris, Ernest. "Canoeing on the Rio Negro," *New York World*, February 8, 1881, p.3.

[29] "Ascending the Rio Negro," *New York World*, February 15, 1881, p.7.
[30] The black caiman, *Melanosuchus niger,* which, along with the American alligator, is one of the largest extant reptiles.
[31] Morris, Ernest. "Ernest Morris's Travels," *New York World*, February 22, 1881, p.3.
[32] Morris, *In Search of Human Heads*, Chapter XIV.
[33] The irritant that caused this reaction is unknown; there are no other documented mentions of the word "cauchy" used in this sense.
[34] "Ernest Morris," *Indianapolis Journal*, April 12, 1881, p.8.

10 HONEYMOON WITH ORCHIDS

After stepping off the ship in New York in late April 1881, Ernest Morris was interviewed by his sponsoring newspaper, the *New York World*:

> "I have about seven hundred plants for Mr. Corning," he said, "that I think will please him, as among them are some that I consider rare—that is they ought to be if there is always as much trouble in getting them as I have had. The most rare is the *Cattleya eldorado*, which grows on very high trees covered with epidendrums, which make them difficult to climb. It is not easy to cut them down, as the wood of the tree on which they grow is very hard." When asked if he intended to return to South America, he said "Yes, indeed. There is no country in the world so beautiful, notwithstanding all reports to the contrary, and the climate when one gets used to it is charming. I leave tomorrow for Albany with the specimens I have gathered, and after a rest at Indianapolis with my family I shall again return to Brazil, as there is yet much to be seen. I have a book in manuscript ready for the printer, which describes some of my experiences in Brazil on my first trip and my trip up the Tapajós river, one of the tributaries of the Amazon. During my last visit, I was unfortunate enough in my hunts for these orchidaceous plants to contract fevers which prevented me from making such trips as I had previously undertaken. After my rest here, however, I shall start out again, and hope to be able to visit Colombia and Ecuador. The [lower] valley of the Amazon is too civilized for me, and I want to get off the beaten path. When I come across an empty beer bottle hung up as an ornament in an Indian hut, it makes me feel as though I was too near home." [1]

While Morris was delivering his plants to Albany, *Scientific American* noted his return, and added a favorable comparison of Morris's achievements to those of Henry Stanley Morton (not mentioned by name, but implied):

> The business of orchid hunting may fairly rank among the most adventurous of the occupations of men, and the number of enthusiastic naturalists engaged in it is larger than is commonly suspected. As a contemporary points out, the owners of great floral establishments in Europe and America keep a regular staff of hardy botanists, who are to them what special correspondents are to a great newspaper. If the truth were known, it would probably be found that professional orchid-hunters have explored more remote parts of the world than the foreign representatives of journals have ever done, but the world at large knows it not, because the orchid-hunters are contented with the discovery of new specimens or filling their wallets and cases with rare specimens, and then returning quietly to their employers; while the special correspondent is bound to write and let everybody know where he is and what he is doing.[2]

Morris, of course, had a foot in both camps, since he was both an orchid-hunter and a special correspondent for the *New York World*.

The *New York World* had been purchased six months before Ernest's return to the United States by railroad baron Jay Gould, who acquired it to counter the critical articles many New York newspapers were printing about Gould's business ethics and character. Moreover, as if to prove his critics right, Gould intended to use the newspaper to print negative items about companies he sought to acquire, driving down their value so that he could grab their stock. Gould likely had little interest in how the rest of the newspaper was run, but was aware of its special correspondent, Ernest Morris, since Gould and his daughter had also recently taken up orchid growing. Morris, at that point, was the only American orchid-hunter familiar with South America.

In 1880, Jay Gould had purchased Lyndhurst, a Tarrytown, New York, country estate with a renowned Gothic Revival house from the widow of businessman George Merritt. A devoted gardener, Merritt had constructed an enormous greenhouse, over 300 feet long, including an aviary tower that reached 100 feet high. Gould immediately refurnished the vacated greenhouse and purchased thousands of plants to restock it. However, in January 1881, the entire vast wooden structure burned down.

Jay Gould, half-length portrait, facing left. [Library of Congress Prints and Photographs Division Washington, D.C. 20540 USA http://hdl.loc.gov/loc.pnp/pp.print. BIOG FILE - Gould, Jay [P&P] (digital file from original item. LC-USZ62-104428 (b&w film copy neg.)]

Gould had the greenhouse rebuilt in early 1881, using frameworks of wrought iron instead of wood. This style had been in use in England for many years, but Gould's new greenhouse was one of the first and finest American examples. Measuring 376 feet by 100 feet, with a central atrium that included a palm room, the Lyndhurst greenhouse was a wonder of garden architecture. Although only one portion of it was devoted to orchids, that one sub-house was large enough to support sections with different climate controls.

While the *Scientific American* article was complimentary, Morris's hometown *Indianapolis Sentinel* was effusive in its praise of him, mixing plaudits of his character with unabashed racism and nationalism. It is important to note these ugly attitudes, because they represented how many in America perceived Native Americans and the manifest destiny of the United States:

> Mr. Morris is still a very young man, and, in appearance, a boy. But in experience, Mr. Morris is a man; he is lithe, wiry, compact, self-reliant and energetic—a born explorer. By profession Mr. Morris is a naturalist, and, in pursuing his profession, an enthusiast. He is possessed of indominable willpower; he never falters or wavers; dangers and trials neither daunt nor discourage him; he possesses exhaustless resources of contentment. He is his own companion. In his hut on the shores of the Amazon or its tributaries, in the trackless forests, in his canoe, on the mountain, or in the valley, he is equally at home. He has the faculty of managing Indians, and bends them easily to his masterly will. He readily masters their language, and, this done, he learns their secrets, and, to the extent desired, takes possession of their country. In describing scenes and scenery, Mr. Morris is always interesting—often graphic. The unadorned facts are full of romantic embellishment. In his descriptions of Indian tribes, their language and habits, of beasts, birds, reptiles and insects; of plants and flowers, fancy and imagination are not required. The naked truth is stronger than fiction, and those who are so fortunate as to hear Mr. Morris describe his travels through tropical South America are instructed and delighted. Indiana should feel great pride in Ernest Morris. He has almost unaided, fought his way to eminence, and is destined to ascend to pre-eminence. His career is a grand example to young men. It is worthy of imitation. We hear much nowadays about self-made men. Mr. Morris is not only a "young man of a thousand," but it would be difficult for the entire population of the country to name a dozen instances of such self-reliance as the career of young Morris brings into prominence. We are glad to know that the youthful explorer is to give, in book form, his experience, explorations, and discoveries in interior South America—confident that it will have a large sale, and while adding to the fame of the author, will be a valuable contribution to the literature of the times.[3]

When Ernest Morris returned to Indianapolis in the middle of May 1881, his ambitions took a back seat to the lingering effects of the countless days and evenings of solitude and loneliness he had experienced in the rainforest. Since the summer of 1877, he had maintained a correspondence with the young woman who was enamored by one of Morris's public lectures, Catherine R. "Kate" Perry. Kate Perry was the daughter of John C. Perry, owner of the oldest grocery wholesale business in Indiana.[4] In November of 1880, while Morris was up the Rio Negro, Kate Perry turned eighteen. When they met again in person after Morris returned to Indianapolis, their rekindled ardor for one another took precedence over any other concerns.

Aerial View of Greenhouse, Looking Northwest - Lyndhurst, Greenhouse, 635 South Broadway, Tarrytown, Westchester County, NY [Library of Congress Prints and Photographs Division Washington, D.C. 20540 USA http://hdl.loc.gov/loc.pnp/pp.print. HABS NY,60-TARY,1B–18 (digital file from original item)

John C. Perry, Kate's father, must have weighed the merits of the match. Morris had less than a thousand dollars to his name even after the sales of his plants, and was engaged in a dangerous occupation that took him to a different hemisphere for long periods of time. He had no house or property of his own. On the other hand, he had achieved national fame as a special correspondent, and universal acclaim for his character and work ethic. Most people were predicting even greater things for the young man; and it was likely that his celebrity and income would only increase once his manuscript was published. The publisher that Morris mentioned had invited him back in 1877 to submit a manuscript was one of the nation's largest: Charles Scribner & Sons. What Morris did not realize was that Scribner's invitation to submit a manuscript had been superseded.

Scribner's Monthly continued to print installments of Herbert Huntington Smith's Amazon travel articles throughout the second half of 1879. By late December of that year, the full-length book version was published by Charles Scribner & Sons under the title *Brazil, the Amazons and the Coast.* It was widely advertised and received glowing reviews. The *New York Herald*'s review of Smith's book even contained some pointed words that seemed to reference Morris's announced plans to publish (and which also served as a dig by the *Herald* against their rival, the *New York World*):

> It was only after another year so spent [in fieldwork] that Mr. Smith began to project the publication of this work, and it is characteristic of his bias for thorough research that instead of throwing together at once the diaries of his finished explorations he began carefully to study the works already existing about Brazil. For the purpose of this publication he made two more trips to Brazil at the expense of Messrs. Scribner, being accompanied on one of them by the artist, Mr. J. Wells Champney, who has furnished over a hundred very graphic illustrations of Brazilian life and scenery...we have here no mere record of ephemeral travel, but a book of permanent value for the student, the best written upon the subject for many years.

Even so, while Herbert H. Smith basked in the success of his book throughout the year 1880 and the first months of 1881, he was also exposed to one column after another appearing in the *New York World* detailing Ernest Morris's adventures on the Upper Amazon, Rio Ucayali, Rio Napo, and Rio Negro. Whether Smith felt that Morris was grabbing too much attention, or whether he just judged it time to get back to scientific fieldwork himself, he made plans for a new expedition to Brazil that would include his wife and a few assistants. In early May 1881 (just a week after Ernest Morris returned from his two-year South American journey), Smith announced he would soon be leaving on an ambitious, multi-year excursion to Brazil. Smith's projected itinerary sounded very familiar: he intended to ascend the Tapajós river, past all the cachoeiras, past the Rio Cururu where Morris had ventured, and up the source tributaries of the Tapajós: The Rio Sao Manuel, the Rio Teles Peres, the Rio Verde and short portage to the Rio Cuiabá, which flowed south to the city of Cuiabá. Smith also had ambitions beyond that arduous task; he told a newspaper interviewer:

> This river [pointing to the Xingu on a map] is now probably the longest unexplored stream in the world. As you see, it is very

> nicely placed on the map, but its real course, as every geographer can tell you, is absolutely unknown. Only a small part, near the mouth, was explored by Prince Adalbert of Prussia. The English [Royal] Geographical Society has long had its heart set on this prize, and probably soon will send someone to explore the river, unless we Americans get ahead of them. For my part, I would readily undertake it, but I fear that I cannot with the funds at my disposal.[5]

Though his opinion of Herbert H. Smith was never recorded, it can be conjectured that Morris viewed him with a mixture of envy and dismissiveness. He knew that Smith was a fine prose writer; had an excellent education; and had been tutored by one of the great American explorers of the Amazon, Charles F. Hartt. However, Morris also knew from the way that Smith had sheltered himself on the Confederado plantation at Santarem that Smith did not mingle with, or learn from, the uncivilized Indians; and had limited toleration for discomforts. Morris probably would have judged that Smith had little chance of ever journeying far up the Tapajós; and, if he read Smith's book, would have criticized it as entirely missing the raw beauty of the forest wilderness and the thrill of venturing off the beaten path. Morris would have felt that Smith did not experience the Amazon with the vibrancy and intimacy that he did.

The wedding of Ernest Morris and Kate Perry took place on August 3, 1881. Afterwards, the couple left immediately for Brooklyn, where Morris prepared an outfit for their honeymoon expedition. Ernest and Kate boarded a boat bound for Pará on August 20, arriving in Brazil in early September. They found that Pará was in the midst of a severe outbreak of yellow fever. Edward S. Rand now made Pará the headquarters of his plant exporting business, but when Ernest went to visit his friend, he discovered that he was seriously ill with yellow fever. The newlyweds made plans to escape Pará as quickly as possible and move up the river to Manaus, which was said to be fever-free. While they waited for a steamer, Morris called on his friend's bedside every day; but when it came time to leave, they feared that Rand would not recover.

Ernest and Kate arrived in Manaus on October 10, 1881. The *New York World* announced that Morris intended to once again attempt to ascend the Rio Japura, an ambition he had held since first visiting Tefé in early 1879. At that time, he had been stopped by warnings about hostile natives, which he heeded only because he lacked trading goods. Kate had no intention to accompany Ernest up the Japura; instead she would stay in

Manaus during the several months that he was away exploring. However, upon arriving in Manaus, Ernest took Kate on a couple of shorter, but still strenuous, orchid-hunting expeditions.

Their first excursion was to Lago Manacapuru, about eighty miles west of Manaus on the Upper Amazon, (Solimões). A steamer took them to the village of Manacapuru, from which they set out by canoe up the Rio Manacapuru to the lake of the same name. The land just to the north of the lake was home to the remaining villages of the unconquered Waimiri-Atroari people–the same warriors who had attacked the village of Moura on the Rio Negro, which Morris had visited the previous year not long after the violence occurred. Morris seemed unconcerned over any danger, as he had been told that the Waimiri-Atroari that lived near the lake were less hostile than those further north in the forests. They camped at an old Indian hut. Morris had a native guide, Manuel, who agreed to show him an area where rare orchids abounded. Morris described his trek with Manuel in a column for the *New York World*, in which he captures the thrill of the orchid-hunter:

> By 9 o'clock we entered a forest, the trees of which were low, wide, and spreading. Great *Bromelias* and *Tillandsias* which had fallen from the trees were growing on the ground, some with spikes two feet long bearing curious red flowers. Now for the first time did I see orchids. Many *Epidendrums* covered the trees, while growing on the decaying stumps were several plants of *Cycnoches*, to me the most curious of all orchids. I passed these by, for I was looking for the plant that Manuel had shown me some days before. Deeper and deeper we entered the damp, silent forest, wet to the skin and tormented by ants; for in cutting our way we frequently brought a nest down upon us. In the afternoon, we arrived at the banks of a large creek, where Manuel pointed out the plants. By the aid of my glass I saw that they were *Cattleyas*, but of what species I could not determine. The leaves were broad and of a darker green than those of *Cattleya eldorado*. During the rest of the afternoon I gathered many plants. Towards evening Manuel wandered off into the forest after *bacaba*[6], and on his return brought me a magnificent plant in bloom. Without a doubt, I had [now gathered] several hundred plants of *Cattleya wagneri*.[7] The flowers are pure white, lip large, also white, and the throat of a deep yellow. On our return to the lake one of the plants bloomed, and I was delighted to find it a variety of *Cattleya wagneri*, for in the white lip was a streak of purple. On the following day we began our return march and reached the Indian

> hut with hardly a rag on us, but satisfied with the result of our days in the forest.[8]

In late December 1881, Morris and Kate took a steamer up the Rio Negro to the little village of Tauapassu, where Morris had been treated kindly by the residents the previous year. At that time (December 1880) Morris had spent Christmas attending a festa and dance held in the village, and had enjoyed himself greatly; now he wanted to share both the charm of the native celebrations and the beauty of the flooded forests with his bride. Kate later described the experience to an interviewer:

> One day I coaxed Ernest to let me go with him, and as it was impossible to go through the woods with skirts on, I donned a pair of his pants, and with a pistol at my belt and knife in hand, we cut our way through bushes and underbrush in quest of plants. I think my husband knows what family a plant belongs to the minute he lays eyes on it. I also went with him a great many miles on the Amazon River [sic, Rio Negro] in a canoe, when the water was up to the tree tops, to attend the Christmas festival at the neighboring village, and was amazed to find the young folks dancing all our new fancy dances in their bare feet, far more gracefully than many of our belles.[9]

However, Kate very quickly tired of the discomforts of camping in the wild, living in run-down thatch huts, and subsisting on salt fish and farina. In Manaus, she also found herself very isolated, since the wives and daughters of the Portuguese descendants were kept shuttered in their homes, and did not engage in any social activities. Though it was never revealed, Kate likely knew little Portuguese or língua geral, and therefore could not converse with anyone other than Ernest or the stray English or American travelers in Manaus that he introduced to her. The *Indianapolis Sentinel* had predicted before their marriage that "the 'boy naturalist' and South American explorer will find it difficult to traverse the jungles with a bride by his side,"[10] but Kate lasted it out in Manaus for four months, with more pluck than some of Earnest's earlier travel companions—and his brother.

Ernest had originally planned to leave on his journey up the Japura on January 1, 1882, accompanied not by Kate, but by a twenty-three-year-old Englishman named Robert Warberton Toulmin. Toulmin had emigrated to Brooklyn in the late 1870s, and, apparently, had met Morris in that city. Enthralled by the stories of Morris's adventures, Toulmin asked to join Morris's Japura expedition. Their plan was to ascend the full length of the Japura into the Andes; and then follow the Magdelena River down to Bogota, Colombia. Toulmin boarded a ship to Brazil in New York on Dec. 1, 1881. What happened to him on that voyage is unknown, as were his whereabouts for the next two years, but he did reappear as a Brooklyn real estate agent in 1883. Toulmin bowed out of the planned expedition at some point after stepping on that boat; neither Kate nor Ernest mentions meeting him in Manaus. More than likely, Toulmin made it as far as Pará, reconsidered, and headed back to Brooklyn before ever steaming up the Amazon.

Ernest was depending to some degree on material support from Toulmin for the expedition up the Japura. He now had to revise his plans for a journey that would require a smaller outfit. Regardless of Ernest's plans, Kate insisted on leaving Manaus on February 10, 1882. She arrived alone in Pará just between the monthly passenger steamers, but was so anxious to depart that she boarded a three-masted schooner as the only woman passenger. Her traveling companions were a monkey and two parrots.[11] She reached Norfolk, Virginia on March 10, 1882, and from there took trains back to Indianapolis. From any perspective, it was an odd way to end a honeymoon, and Ernest Morris could not have helped but to feel remorse after she left.

Morris considered taking up the expedition that he had been discouraged from making two years earlier: taking a steamer to Iquitos, Peru; then another steamer down the Huallaga to the town of Yurimaguas. From there he could hire a canoe to ascend the Paranapura River to the village of Balsapuerto. A hike of ten days through valleys in the Andes would get him to Moyobamba, Peru's prime orchid-growing region. After collecting specimens there, roads and rail could get him to the coast and down to Lima, where he could catch a ship bound for Panama. However, Peru was still embroiled in the War of the Pacific. Chilean troops had captured Lima and were waging battles in the interior of the country against Peruvian guerilla forces. Travel in Peru would not only be dangerous, but there were also no guarantees that any steamers or railroads would still be running.

In the end, Morris may have realized that his new marriage might be in peril if he stayed in Brazil any longer. In early March 1882--just one month after Kate had left--Ernest boarded at steamer at Pará with 1800 orchids, three live boa constrictors, and "other curiosities" and set sail for New York. When he arrived, he explained to reporters that the number of orchids he had already collected required him to curtail plans for any further explorations. About a dozen of his orchid specimens had been shipped forward to Germany, where expert botanists there were to determine if they were new species.[12] It is likely that the specific person these were addressed to was Heinrich Gustav Reichenbach, the world's foremost expert on orchid botany.

Ernest reunited with Kate at the Indianapolis home of her father, John C. Perry. As yet the young couple did not have a place of their own. If Ernest was repentant about having sent Kate back to the United states alone, he hid it well. A reporter from the *Indianapolis Journal* asked him if he planned to return to South America:

> 'I have not yet decided whether I will or not, but for the next few weeks I will remain at home, devoting my time principally to putting my notes into manuscript. I am making arrangements for the publication of two volumes, one of which is in manuscript, and the other in preparation. My determination in the matter [of a future expedition] will be reached that quick,' he said, snapping his fingers. 'But at present I cannot say what I shall do. I need rest, and I intend to take it now. Besides, my friends here are urging me to remain.'[13]

Earlier, Kate Perry Morris had been asked by the *Journal* about her husband's plans once he returned. She replied, "He will then prepare a book to be published by Scribner of New York under the title of *A Trip Up the Tapajós In Search of Human Heads.*"[14]

This was the first indication that the publisher that Morris had been negotiating with to publish his Tapajós adventures was Charles Scribner & Sons. Those talks likely began immediately after the fanfare that attended Ernest Morris's arrival in New York with ten dried human heads in 1877. However, in the meantime, Scribner had received and published a much more polished, scholarly, and wonderfully illustrated book on Brazil, Herbert H. Smith's *Brazil, the Amazons and the Coast.* When Ernest T. Morris eventually presented his manuscript to Scribner, the editors were likely not only disinclined to publish another work by an Amazon naturalist so soon—they were also disappointed at quality of the manuscript that Morris delivered.

The text of Morris's *In Search of Human Heads,*[15] as he presented it to publishers, had many structural flaws, most of which could have been overcome with a good editor and a willingness on Ernest's part to rework the text. The text is a somewhat disjointed collection of fourteen chapters. The first three chapters are based on Morris's notes from his first journey to the Amazon when he was eighteen, and is written in the style of other travel narratives by naturalists, such as Henry Walter Bates's *A Naturalist on The River Amazons*. Morris uses the first-person plural to refer to himself, i.e. "we," "our," etc., which was often used by scientific observers of the time. Those chapters are unexceptional, though they plainly show the author's love of the tropical wild; his keen eye and knowledge of flora and fauna; and his willingness to endure discomforts.

Starting in the fourth chapter, Morris begins the story of his second trip to Brazil, where he ascended the Tapajós and Rio Cururu rivers and traded with the Campineiros group of Munduruku Indians for eleven dried, smoked human heads. This section, which continues through the eleventh chapter, is a compelling true-life adventure story, marred only by Morris's continued use of the first-person plural. Chapter Twelve interrupts the narrative of his journey by presenting an essay on the Munduruku: their rituals, dress, villages, artifacts, tattooing, food, population size, etc. The thirteenth chapter resumes with his return down the Rio Cururu and Tapajós rivers, ending at the town of Santarem, where the fourth chapter started. The fourteenth, final chapter is a guide for tourists: a laundry list of all the poor accommodations, bad food, and insect pests that a traveler to the Amazon could expect to encounter. Morris must have meant to discourage all but the hardiest souls to attempt such a trip.

It is difficult to know what Morris intended the book to be: a true adventure tale; or a naturalist's travel narrative; or an ethnological study; or a guide for visiting Americans. Many passages within it match a style that Morris exhibited in all his newspaper writings: in nearly every column and letter, he refers to someone demonstrating foolish human foibles; and he mentions the physical discomforts and ailments that plagued his travels; but he also always refers to some natural beauty that inspired awe and dwarfed all other troubles. This stylistic pattern worked in his newspaper columns and letters; but extending that formula to book-length was beyond his talents, without some help. An editor would have advised him to stick to the adventure story; condense the first four chapters into one; move the Munduruku essay to the end as an appendix; and insert notes on all the nasty creatures found in the final, travel-advisory-chapter, into the adventure story.

Even had the manuscript been improved, Charles Scribner & Sons and other major publishers of the early 1880s would have viewed it as a narrative of a prolonged exposure to savages: living as primitively as they did, and bringing back examples of their gruesome, barbaric war trophies.

It was a topic they might have judged more suited for dime novels. Morris descriptions were too unpleasantly graphic: the insect plagues, the disgusting foods, the noisome smells, the abysmal hygiene, the shameless nakedness, the lack of spirituality, etc. would strike many 1880s readers as shocking and amoral. Moreover, Morris several times insisted that the uncivilized natives were more capable, skilled, and happier than those touched by civilization; and that the descendants of the Europeans (Portuguese) were lazy, stupid, and overprivileged.

In the United States of the early 1880s, the major battles of the Indian Wars had recently ended (although the Wounded Knee Massacre was yet to come) with the subjugation of most native nations and their constraint to reservations. On those reservations, their traditional way of life was discouraged and forcefully repressed. Popular culture would later romanticize "the noble savage" and mourn the inevitable passing of "the Vanishing American," but in the early 1880s, the public sentiment of most white Americans was that barbaric aborigines, in the United States and elsewhere, needed to be assimilated or eradicated if they stood in the way of "progress." Had they seen Morris's manuscript, many might have thought it was promoting the exact opposite.

Ernest Morris received a rejection from Charles Scribner & Sons on July 18, 1881, just days before his marriage to Kate Perry.[16] According to an interview she gave after his death, Ernest submitted his book to other publishers for years afterward, without success. Limited to the money he had earned from selling the 1800 plants to Erastus Corning, Morris purchased a modest home at 667 North Delaware Street, Indianapolis. That left him without any savings. By the autumn of 1882, he was working for his father-in-law, John C. "Jack" Perry, in the wholesale grocery business. His duties were at least tangentially related to his knowledge and skills—he was charged with buying and selling shipments of fruit, including imports. In late 1882, Kate announced that she was pregnant. A girl, Enid Perry Morris, was born to the couple in June 1883.

Soon after their child's birth, the Morrises visited with Ernest's father, Sanford Morris, where he now resided in Shelbyville, Indiana. The local paper, the *Shelbyville Democrat,* noted their presence in town, and offered a prediction:

> The spirit of the explorer still possesses the "Boy Naturalist," and it would not surprise us to hear at any time that he had started for South America for another trip. His charming wife and handsome three-month-old boy [sic, girl], who came down with

> him yesterday, will probably, however, prove a stronger attraction than any possessed by that wonderful country where he acquired that fame.[17]

The *Shelbyville Democrat* was wrong.

---

[1] "Rare Orchids from Brazil," *Indianapolis Journal,* May 2, 1881, p.5.
[2] "Orchid Hunting Along the Rio Negro," *Scientific American,* May 7, 1881, p. 292.
[3] "Ernest Morris," *Indianapolis Sentinel,* May 21, 1881.
[4] A biographical entry for John C. Perry can be found in: Dunn, Jacob Piatt, and G. W. H. Kemper. 1919. Indiana and Indianans: a history of aboriginal and territorial Indiana and the century of statehood. Chicago: American Historical Society. Vol. 4, p. 1766.
[5] "Explorations in Brazil," *Utica Morning Herald,* April 7, 1881.
[6] *Oenocarpus bacaba, a fruiting palm*
[7] The white-flowered form of *Cattleya mossiae*
[8] "Wanderings in Brazil," *New York World,* February 28, 1882, p. 7.
[9] "Brave Little Woman," *Shelbyville Daily Evening Democrat,* March 24, 1882, p. 1.
[10] "An Exchange," *Indianapolis Sentinel,* July 16,1881, p.4.
[11] "Relates Thrilling Honeymoon into Land of Headhunters," *Indianapolis Sunday Star,* July 4, 1926, p.32.
[12] "The Young Explorer," *Indianapolis Journal,* April 19, 1882, p.8.
[13] Ibid.
[14] "An Interesting Talk," *Indianapolis Journal,* March 24, 1882, p.8.
[15] The only copy of the custom-bound draft has this shortened version of the title, dropping "A Trip Up the Tapajós."
[16] Series 22: Rejection Files, 1873-1987. *Archives of Charles Scribner's Sons.* Department of Rare Books and Special Collections, Princeton University, Princeton, NJ.
[17] "Local News," *Shelbyville Daily Evening Democrat,* September 20, 1883.

11 THE RESTLESS TRAVELER

The terms of Ernest Morris's status as a special correspondent with the *New York World* were never disclosed. He could have signed one or several written contracts; or a per-column payment. Once Morris began writing for the *World* in 1879, he stopped sending letters addressed to his hometown newspapers, the *Indianapolis Journal* and the *Indianapolis Sentinel*. On one occasion, he remarked that his deal with the *New York World* precluded him from writing for other papers. During his honeymoon expedition of 1881-1882, four of his columns appeared in the *World*; but on his return to the United States, Morris claimed that he had sent fourteen letters to the newspapers, using unreliable locals and rubber traders to convey them to the Brazilian postal service. He also mentioned that he needed to write some columns before considering what he might do next, implying that he owed the *World* some pieces.

However, after the spring of 1882, the *New York World* stopped publishing anything written by Morris. The newspaper was losing readers and money. Jay Gould proved to be an uninterested owner; and too many people had seen through Gould's ploy of publishing slanted articles to degrade the stock value of companies he aimed to acquire. In April 1883, Gould sold the *New York World* to Joseph Pulitzer, who quickly made sweeping changes in the paper's editorial direction and tone. If Morris approached Pulitzer's new regime at the *New York World* in the hopes of obtaining support for a new expedition, he came away disappointed. But it appears that Pulitzer and the *World,* although they no longer wanted to use Morris as a columnist, also did not want him to be recruited by a competitor—and pointed to a contract that bound his freedom to do so.

In 1883, while working for his father-in-law in the grocery wholesale business, Morris met a twenty-four-year-old man named George Bowles. Bowles was the secretary and spokesman for Thomas C. Moore, a railway agent responsible for coordinating all east-bound traffic from Chicago. Bowles was easy-going, garrulous, and loved to travel. He had many friends and contacts among Chicago's newspaper reporters. He was also a fan of theater, and romanced several actresses, but at the time he encountered Morris was still single. He would eventually marry twice, both times to actresses.

Morris and Bowles met while reviewing railway freight rates, but developed a rapport when Morris revealed his history as an explorer. Bowles was fascinated by his tales of Amazon travel. The two men soon developed a plan: since Morris was no longer supported by the *World* and was blocked from writing for another newspaper, Bowles would approach one of the Chicago newspapers and convince them to make Bowles their Special Correspondent during a trip to South America accompanying Morris. Morris would provide all the natural history observations that Bowles would need, and Bowles would simply publish it all under his name.

Bowles resigned from his position working for the railroad agent Moore and announced that he was embarking on a two-year journey to the tropics. For his part, Morris told Indianapolis newspapers that he would be gone for a year; and that his plant-collecting route would take him from Georgetown, British Guiana, down the Essequibo River, and from there he would get to the Rio Negro.[1] How he intended to do that is unknown, although a tributary of the Essequibo runs to within a twenty-mile portage to the Rio Branco, which flows into the Rio Negro. Bowles and Morris departed for their journey in mid-February 1884. Their first destination was Port of Spain on the island of Trinidad, a British colony just off the coast of Venezuela.

Bowles' first *Chicago Tribune* column, published after a delay of several months, painted a favorable picture of Trinidad. His writing style was dramatically different from that of Ernest Morris. Bowles used the first-person singular voice; and the only discomforts he noticed in Trinidad were that the hotel buildings all looked the same; the coffee was strong; and the bread was oddly shaped. Once Bowles and Morris left Trinidad and arrived at Georgetown, British Guiana, on the coastline known as Demerara, Bowles opinions turned decidedly worse. He was surprised at the mix of nationalities, for it appeared that Great Britain had imported cheap labor from around the world--particularly India--to support their sugar and rum industries. Slavery was outlawed, but indenture abounded. As for the British colonists, Bowles noted:

> The white population is largely composed of young Englishmen, who, being of little account at home, for want of brains, or for other reasons, have been sent out to this fever-bed to be got rid of. And they are apparently doing all in their power to further their friends' wishes in this respect; for from early morning till far into the night there is an incessant gulping down of liquor, and nothing but whisky, gin or brandy seems to be called for. No longer does the good American hear the cry, "Two beers," but in its place comes the languid drawl, "Waitah, bwing me a gin swizzle, aw, and a bwandy straight." I cannot dwell upon the other charge against these Demerarians, but will only quote the remark of an old sea captain, who said to me, briefly but truly, "Guiana is another Sodom."[2]

Bowles and Morris did not tarry long in Georgetown, waiting only for the twice-weekly steamer that ascended the Essequibo about fifty miles to the point where the Cuyuni and Mazaruni Rivers join the Essequibo. At that junction, there was a small village, Bartica; and, a few miles up the Mazaruni, the colonial Penal Settlement. The prison and the missionary in Bartica represented the last outposts of "civilization" in the interior of Guyana. In Bartica, the two Americans were greeted by the padre and curious Carib Indians, naked except for loincloths. Bowles was impressed by how fit and muscular the natives appeared. Several of the men carried very large bows—six to seven feet in length, with long arrows; while others had blowguns. Bowles and Morris were shown to their guest accommodations—a three-sided thatch hut.

From Bartica, Bowles and Morris ventured on a side trip up the Cuyuni River, accompanied by six native paddlers: three Arawak, one Carib, and two Akawaio Indians. The Akawaio assumed the roles of bowman and steersman, since they knew the Cuyuni waters. Ascending the Cuyuni, Bowles got a taste of traversing rapids that Morris knew so well: tying ropes to the bow, carrying the rope forward to a tree or rock, and pulling the canoe forward. As on the Tapajós in Brazil, each cascade on the Cuyuni had a name: Seregatava, Tivruni, Ematuba, and Kamaria were ones they passed. After camping one night, they reached Wariri Creek, the site of an abandoned British gold mine, the Wariri Mine. They observed orchids on the banks, but none particularly rare.

Camping in the wild also gave Bowles a sample of the night noises of the rainforest that Morris enjoyed: the frogs, goatsuckers (nightjars), jaguars, and howler monkeys. Bowles found it impossible to sleep. At the mine, they found quartz infused with gold; also a few specks of gold while panning at the mouth of a nearby creek. They discovered they were not alone; they met three French prospectors employing a "long-tom" rocker to separate gold, a technique reminiscent of 1849 Gold Rush miners. Morris and Bowles stayed on the Cuyuni for a few days, then pointed the canoe back downstream forty miles and reached Bartica in one day.

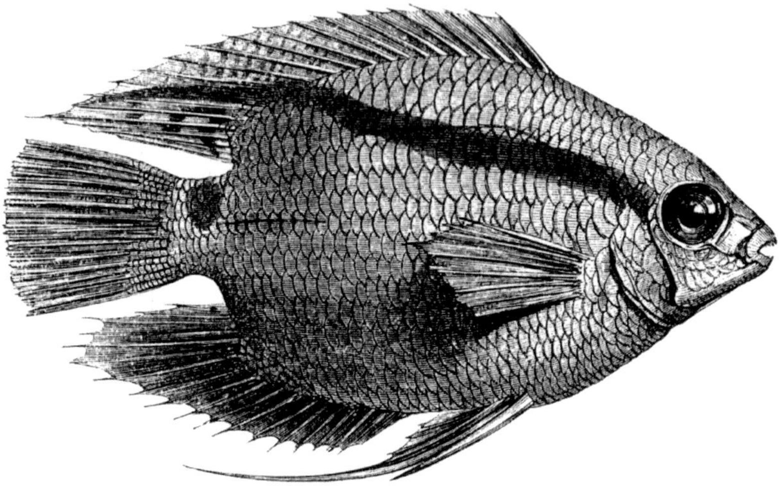

Acará (*Mesonauta insignis*). From: Bates, Henry Walter. *The Naturalist on the River Amazons.* London: John Murray, 1863. [Public Domain]

Their next journey took them thirty miles further south, ascending the Essequibo, to the Aretaka rapids. Their Carib guides brought them to a village near the rapids. While there, they were invited to accompany a hunting party searching out a pack of peccaries, an event which ended with everyone, including Bowles, scratched up and tattered from the run through the forest. The hunt was successful, and resulted in a feast, and stories told round the fires of wood demons and the Carib creation story. According to that story, the Carib ancestors once lived in the sky, but discovered a hole that looked down on the earth. Through the hole, they saw the earth was chaotic and dirty, and so some among them volunteered to jump through the hole to go down and put things in order. After many had jumped, a fat woman dropped into the hole and got stuck, trapping the ones that had already jumped on the earth. Those thus stranded founded the Carib people.

The Carib village at Aretaka represents the last known event in the expedition of Morris and Bowles to British Guiana. From Bartica, Morris returned to Georgetown and boarded the British bark *St. Lawrence*, a cargo vessel carrying a load of sugar, bound for Philadelphia. He arrived on April 18, 1884, having spent just one month in the colony.[3] The only explanation Morris gave was that upon his arrival in Georgetown he "found the market overstocked with collections." This could mean that the English patron for whom he had planned to collect had found plants from another source; or that British Guiana had been overrun by orchid collectors. The *Indianapolis News* of April 21, 1884, noted on Ernest's return: "He brought back a stock of plants for Bruce, the florist of this city. And he will make another trip in a few months."[4] Whether Bowles returned with Morris or lingered in the Caribbean is unknown; he surfaced later that spring in Chicago, saw five of his columns published in the *Chicago Tribune*, and resumed his position as secretary to the rail agent.[5]

Ascending the Rapids.

Ascending the Rapids by James Wells Champney. From: Smith, Herbert H. *Brazil, The Amazons and the Coast.* New York: Charles Scribner & Sons, 1879. [Public Domain]

There are no clues that suggest a different explanation for Morris's return other than the one Morris offered: the supply of orchid plants was larger than the demand, and Morris had been counting on a stronger market to fund a lengthier expedition. It could be that he learned this shortly after arriving in Georgetown; and only ventured down the Essequibo to satisfy George Bowles' desire to see the wild tropical forest. In early May 1884, an Indianapolis paper noted that Morris looked robust, so there is nothing to intimate that illness caused his return.[6]

During the short period between Morris's first and last trips to South America, 1876-1884, the collection of orchids from the wild had become a large-scale industrial enterprise, thanks to a handful of British orchid growers: Dr. John Lowe; Beechers; Veitch & Sons; and the "Orchid King," Frederick Sander. In the 1880s, Sander had as many as twenty-three men scouring the world for elusive plants. He had sixty greenhouses at his main facility at St. Albans, near London; and had branches in Belgium and (briefly) in New Jersey in the United States. The St. Albans conservatories were said to house between one and two million plants. The rivalry among these competitors was so fierce that the hunters would clean out an orchid area, just to leave nothing for other hunters; or would sabotage the plants their rival had collected; or would bid against one another for the services of native guides.

Sander's chief orchid-hunter was Benedikt Roezl, who bragged about clear-cutting orchid habitats in South America after he had taken all he could. Many of Sander's hunters met violent deaths, including Roezl's nephew, Edward Klaboch. These professional orchid-hunters shipped home huge volumes of plants, resulting in some governments setting prohibitions and stiff tariffs against the mass removal of plants. Between the efforts of the British hunter and those of the native orchid-hunters hired by Morris's friend Edward S. Rand, the relatively small quantities that Morris could collect on his own would not have been worth the effort, since the British and Rand could sell plants at lower prices.

When he got back to Indiana, Morris gave a series of lectures around the state, including a talk at Purdue University; but it appears he gave these with little planning, and was traveling from town to town as a salesman.[7] In August 1884, he left the employ of his father-in-law and signed on as a salesman for Snyder's Preserves of Cincinnati, though he and Kate still resided in Indianapolis.[8] That job lasted less than a year. In July 1885, the Morris family moved to Keokuk, Iowa, where Ernest covered territory as a salesman for Shaw & Thomas, spice dealers headquartered in New York City.[9]

While employed by Shaw & Thomas, Morris convinced them to send him on a business trip to New Orleans to attend the North, Central, and South American Exposition in December 1885. En route to Louisiana, he stopped by the town where he was born, Georgetown, Texas; and from there traveled to Galveston, to visit a distant older cousin, Noah Noble John, a prominent physician and member of the city's Board of Health.[10] John had survived and treated hundreds during the horrific yellow fever epidemic of the 1860s. Galveston, perhaps more than any other American city, had experienced numerous outbreaks of mosquito-borne diseases. While Morris's stop at Galveston could have just been a family visit, it may also be conjectured that Morris was consulting John about his lingering fevers.

The Exposition in New Orleans featured a Horticultural Hall with hundreds of displays of plants and fruits from all the Americas. The building featured glass conservatories around its perimeter, in which flowering plants, cacti, bushes, small trees, and ferns were placed. In the southeast corner, a 250-foot long hothouse contained hundreds of *Tillandsia,* the air-growing plants so familiar to Morris, as he saw them growing from the trees in the rainforest. It must have been a bittersweet experience for him to see them in such an artificial setting. One report mentioned that Morris spoke at the Exposition, but the subject of his talk is not known.

The Morrises's stay in Keokuk was short. Kate Perry Morris became very ill and returned to Indianapolis with their daughter Enid to recuperate. Ernest moved on to yet another traveling sales job, this time for the grocery wholesaler Isaac Stadden of Fort Scott, Kansas. He relocated to Kansas while Kate and Enid remained in Indiana, as Kate was still not in good health.[11] Ernest set up a window of his Amazon artifacts in Haven's drugstore in Fort Scott, perhaps in exchange for medicine for his bronchitis.[12] Two artifacts that gained prominent attention were the skins of a boa constrictor and an anaconda.

Morris had learned over the years that people in the United States always assumed the Amazon contained vast numbers of big, deadly snakes that were a peril to travelers. When asked, he dispelled that notion by first pointing out that snakes were rarely seen in villages or on traveled paths, or even in dry woods; they preferred the thick rainforest or around pools, streams, and marshes. Morris had never been bothered by the large constrictors; and the skins he obtained had been brought to him by villagers.

However, Morris had many narrow escapes with snakes while orchid hunting. The snakes he feared were smaller, poisonous serpents lying unseen on the forest floor—or worse yet, in tree branches as he reached out to grab a plant. Though he never admitted to being bitten himself, Morris was familiar with: coral snakes, of which there many species; *jararaca* (*Bothrops jararaca)* vipers; and the bushmaster or surucucú, another pit viper. In Morris's era, few treatments existed for snake venom; he once stated that the most common treatment was to overcome the poison by imbibing alcohol.

Le Souroucoucou by Édouard Riou. From: Biard, François. *Deux anness au Brésil.* Paris: Librairie de L. Hachette, 1862. [Public Domain]

Out of nostalgia, Ernest penned a letter to the editor of the *Fort Scott Monitor* that represents his last public writing that appeared in print:

> "Sunny Kansas," at least I have found it so, during my short residence, but sitting alone tonight gazing into the cheerful fire I see a sunnier land, a land of palms and mighty rivers and I can almost hear the rushing sound of the Amazon, and in imagination wander again in those wonderful forest, or drift idly down some lonely stream, listening to the click of my Indian's paddle, inhaling the delicious perfume from the woods, and I almost seem to live over again my long wandering life—the life of a naturalist, not without its hardships or dangers, camping where night overtakes him, slinging his hammock in the trees, and studying nature in nature's own storehouse. Such a life was mine. In looking over my notes my eye catches the date of January 29; below, the heading, in search of *Victoria regia.* What a trip that was and how enjoyed. It was on a glorious January morning, not snow or ice, or biting winds; but warm breezes, cloudless sky, the air redolent with the perfume of flowers, and resonant with the song of birds.
>
> We were three—two of my companions being Indians, who

knew the forest for many miles around. Our object was to obtain that beautiful aquatic plant, *Victoria regia.* The sun had not risen above the trees when we left our little palm hut, armed with guns, the Indians carrying great baskets on their backs, full of eatables. As we neared the river, tall grass grew on either side of the path, every blade dripping with dew, and in a few moments sufficed to wet us thoroughly from the knees downward. It is now lighter. The Amazon seemed a broad lake, not a breath of air to stir its surface, and yet, there comes to our ears, the rushing sound of the vast volume of water, which restless, like the sea, chafes its confining banks day by day, wearing them away with its mighty current, swift as a mill race and resistless as the waves of oceans.

We find our boat and after storing our goods under the little palm cabin, we light long cigarettes, take our paddles, and silently yet rapidly move upstream. There is oppression in the early morn; the dark, somber forest, the broad lonely river, seem to act as a pall on our spirits, for the magnitude of nature is deeply impressed on us. But as the sun arose from behind the lofty trees our tongues loosened, and we laughed and talked, I enjoying the strange stories of my Indians. By nine o'clock we had made many miles when we ran out boat ashore, and built a fire, while one of the Indians went into the forest in search of game. He soon returned, bringing a small titi monkey, which was quickly roasted; it, together with dry salt fish and a farinha, made a most capital breakfast.

A breeze soon sprang up, and hoisting our sail we sped rapidly up the river. We laid aside the paddles and my Indians lay basking in the sun while I watched the ever-changing forest. In the afternoon we left the Amazon and entered one of the numerous outlets of the chain of lakes which border on the Amazon. The stream was narrow; great trees towered far above us. *Epiphytes* hangs from every limb and *Tillandsias* with their great flaming red flowers nestled in the forks of the trees. Orchids were numerous, while palms of many species were to be seen. Dwarf palms grew in wild profusion, some smooth with rich silvery foliage, others armed with long spines. Plants that have their home in the trees and send down great masses of roots; the smaller trees present foliage of every possible shade of green, berries of every color (but none edible) are plenty, and the ground is a mass of delicate ferns and graceful *Lycopodia.*

By night we had not reached the lake, and were forced to camp in the forest. Cutting some of the great palm leaves, we soon had a shelter. Fire made and supper over, we laid around the fire listening to the strange noises that come from the dark forest: now like the barking of dogs; now a moan dying away in a low, sobbing wail; sounds no one can account for, but which lend the weird-like charm to the tropical forest. To sleep—but sleep was impossible, for the mosquitoes descended upon us by the millions. What the mosquitoes had left of us the next morning was early astir, and after several hours of rapid paddling we entered the lake. The waters were clear and cool, the shore densely fringed with palms. Near the inlet we stopped at an Indian's hut. Clean and neat it was, and we were received by pretty girls wearing in their hair the sweet, starry jessamine. After I had answered as to who I was, where from, whether married or single, etc., a mat was laid on the floor and breakfast, consisting of fresh fish and game was served, which we ate with our fingers in a manner which would have excited the envy of a dyspeptic.

Leaving the hut, we again took our paddles and resumed our voyage, stopping often to gather some rare orchid or admire some curious flower; rounding a point before us in a small bay were the plants we sought, great masses of them, many hundred feet in diameter. Those who have only seen this plant as grown in cultivation, grand as it is, can form no idea of its magnitude and grandeur in its native waters. The young leaves, before unfolding, present a solid mass of stout conical prickles; the leaf is peltate, circular in outline, and varies from three to six feet in diameter. The edge is turned up some two inches or more and the leaves look like shallow floating trays. The upper surface of the leaf is a rich green color, the under deep purple, and is traversed by prominent nerves radiating from the center to the margins; there are also many smaller nerves so that the whole under surface of the leaf is divided into quadrangular spaces. These nerves contain air channels and are covered with sharp spines. The flower bud (before expansion) is pear-shaped and the calyx is covered with prickles. The petals are numerous, pure white stamens, many rose colored. The flowers when fully expanded are about a foot in diameter, but vary greatly in different plants. The second day the petals change to rose color, and later to deep reddish purple, which circumstance has caused many to describe two varieties. The flowers, like some orchids, emit their rich spicy fragrance in puffs.

> We gathered many of these flowers, and for one who could bring himself to endure some exposure, and who for a time would himself forego all luxuries in life, and, perhaps, many things which are considered necessities also, no more charming expedition could be imagined than an excursion of a month or two upon these lakes. There is a constant variety of scenery, and, to the lover of nature, each scene would present new attractions. The luxuriant vegetation, the trees and flowers, the abundance of animal life, both in the woods and waters, the pure cool air, the balmy nights, with the breezes loaded with fragrance, would present ever recurring charms.
>
> We remained for over a month on the lakes, everywhere hospitably received by the Indians, when we again turned our faces toward our home, and having the current in our favor, soon reach the Amazon. Borne along by its rapid current, the following day we reached our little village of Juruti, only to make preparations for a long journey into the interior. Where are my companions, faithful to me for many years? João lies buried in the forest he loved so well; Manuel's grave is near the shadow of the Andes–and I stop writing, look at the fire which has died away to embers, and fall to thinking–dreaming that I am once more in the forests, for to me the beauties are so great that a life in the tropics seems to me far more desirable than a life even if it be spent in "sunny Kansas."[13]

Not surprisingly, after penning this public letter, Ernest lasted only a few months longer working for grocery wholesaler Isaac Stadden in Fort Scott. There is no accurate word in English for the melancholy that Morris experienced. The German language has a word, *fernweh,* that is used to express "farsickness" as opposed to homesickness. However much he was drawn to venture again to the Amazon, that urge was offset by many considerations: the moral responsibility to his family; his lack of finances; his bouts of ill health; the lack of patrons; and a dwindling of the market for flora and fauna specimens. By the end of 1887, Ernest was back with his wife and child in Indianapolis, working once again for his father-in-law, Jack Perry.[14]

Ernest Morris made one more trip in pursuit of his passion. In February 1888, he attended the Second Annual Orchid Exhibition managed by the floriculturists Siebrecht and Wadley, held at the Eden Musée amusement center in New York City. It was an event supported by some of the wealthiest society patrons in New York: the Vanderbilts,

Sloans, Webbs, Skidmores, Scribners, Lorillards, Roosevelts, and others. Ernest kept the program of the show among his journals and letters, and had in the pages listing the names of the displayed orchids, he made marks. Some were mentioned in his writings, so it can be assumed that he had collected examples of all the marked species—perhaps even some of the very same plants he had brought back from Brazil:

Cattleya speciosissima ernesti[15]
Cattleya eldorado
Cattleya wagneri
Cattleya trinae corningii
Laelia bella
Oncidium sarcodes
Schomburgkia crispa
Scuticaria steelei
Sophronitis coccinea [*Cattleya coccinea*]
Zygopetalum intermedium
Rodriguezia secunda

It can only be guessed what else Morris might have done while in New York City. Did he visit the American Museum of Natural History, to see one of his Munduruku heads on display? Did he drop in on any more publishing houses, still trying to get his book in print? Did he visit the docks, where steamers to South America arrived and departed? Both his New Orleans trip and the trip to New York City seemed to be driven more by nostalgia than necessity.

As if to confirm the Morris was reflecting on the past and worrying about his future, in 1889 he enrolled in the recently formed Order of United Commercial Travelers of America. This organization was founded to provide traveling salesmen with financial services, such as disability payments or death benefits for their families. By the next year, 1890, Morris was no longer even on the road; he was serving as a secretary to his father-in-law, keeping the account books and writing letters.[16]

Ernest T. Morris died at his home on North Delaware Street in Indianapolis on April 29, 1891, at age 34. His death notice said the cause was pneumonia, complicated by congested lungs; one of the Indianapolis newspapers used the term consumption, which often meant tuberculosis.[17] His obituary was printed in New York, Washington, and Chicago newspapers as well as in Indiana; all of which reminded readers that he had achieved fame as the "Boy Naturalist," "Boy Explorer," and "Boy Traveler."[18] He was buried at Crown Hill Cemetery in Indianapolis, not far from the plot where one of his cousins by marriage, President

Benjamin Harrison, would later be buried. Ernest's grave, like other Perry family members in the same plot, is unmarked. The lack of headstones is an unexplained mystery; perhaps the patriarch, John C. Perry, had planned a small mausoleum but later could not provide the funding.[19] The anonymous resting place of Ernest T. Morris is emblematic of his quick fade into obscurity.

Sanford Morris, Ernest's father, outlived his son by eight years. Sanford's last twenty years were spent in Shelbyville, Indiana. His obituary portrayed a man of Christian spirit:

> After Mr. Morris began his work in Shelbyville, not even a pauper was buried without a service being held. He was the means of many children being placed in the public schools. When the Orphans Home was built, Mr. and Mrs. Morris were placed in charge of it. Mrs. Morris was a teacher of experience. With the aid of Mr. Morris this house has been made a nursery, a schoolhouse, a chapel, and a home. He will be sadly missed.[20]

Following his son's death, Sanford Morris maintained a scrapbook that contained all the clippings related to Ernest's travels and his columns for the *New York World*. In the front of the scrapbook is an inscription: "This book never to be loaned—as next to my Bible I prize it. Sanford Morris."[21]

Edward Sprague Rand, the fugitive orchid expert of Boston, remained in Brazil and became the largest orchid grower and exporter of that nation. His collection grew to over 30,000 plants, said to be the third largest in the world.[22] He was the first to recognize several orchid species and varieties as unique; and they now bear his name as part of their scientific classification. Rand continued to write articles for American gardening magazines throughout his self-imposed exile; and wrote a book of Tapuya folktales, likely derived from the oral traditions of several Amazon tribes. Though he never returned to Boston, Rand did make several trips to New York to sell plants. He became ill in 1897, and decided to hasten his demise with a lethal dose of strychnine.[23] At his death at age 63, he left instructions for his huge collection to be sold in England. It is not known if he ever contacted his estranged family.

The wife and children that Rand had abandoned in 1877 were not left destitute, though they did have to suffer the shame associated with his crimes. Rand's father, Edward Sprague Rand Sr., had stepped in and settled most of his son's heaviest debts shortly after he left America. Rand

Jr.'s two sons, Henry and Edward, were in their late teens when their father escaped to Brazil. Their Dedham estate was sold to satisfy creditors, but they remained in the Boston area, relocating to Cambridge. Rand's son Edward was born as Edward Sprague Rand III, but after his father's disgrace, changed his name to Edward Lothrop Rand. He went on to graduate from Harvard with a law degree, and also followed in his father's footsteps as an amateur botanist. He authored the definitive work on the flora of Mount Desert Island, Maine, where the family maintained a vacation cottage. Rand Jr.'s other son, Henry, did not go into law, but was content to follow a career in accounting. His passion was photography, and he was a leading member of a noted club of photographic artists, the "Old Cambridge Photographic Club." Henry Rand took hundreds of photographs of the landscapes around Mount Desert Island, which are currently being digitized by the Southwest Harbor Public Library.

During his self-imposed exile in Brazil, Edward S. Rand Jr. heard the heartbreaking news of the death of his parents. His father, Edward S. Rand Sr., had spent much of his own fortune settling the debts of his son. In January 1884, Rand Sr. and his wife boarded the *City of Columbus* steamer bound for Savannah, Georgia. The ship hit dangerous rocks shortly after leaving port. The wreck resulted in 124 deaths, with only a score of survivors.

Erastus Corning Jr., Ernest Morris's wealthy orchid-collecting patron, died in August 1897. It is known that he died near his country estate, and rumors surfaced suggesting that he had suffered a heart attack while driving one of his prize trotting horses. It was said that over the years he had spent $1,000,000 on collecting and caring for his orchids. A month after his demise, the *New York Herald* ran an article headlined "Vanished, a Fortune: How a Vast Estate Was Dissipated in a Quarter of a Century." It revealed shocking facts that came to light when Corning's estate was settled:

> Twenty-five years ago Erastus Corning Jr. came into possession of $12,000,000. He died intestate a month ago, leaving not enough to raise a tax on...At the time of his father's death he was forty-five years old. Into the hands of this mature man fell $12,000,000. The fads seem to break loose then. In the greenhouse, the display of orchids grew. From every clime came priceless specimens, whose perfume and color and form this strange man loved above all else...
>
> ...Wise old politicians will tell you that Erastus Corning

> might have been President of the United States. He at least had a grand chance to be Governor of New York. In the convention of 1882, say these wise old politicians, there were more delegates pledged to Erastus Corning Jr. than to any other possible candidate. He had allowed his name to be used, it is said. Then, at the last moment, he withdrew. Something else attracted him—perhaps he wanted more time for his orchids and for entertaining his friends. So Grover Cleveland was nominated, and swept into office triumphantly.[24]

The *New York Herald* writer's guess was probably accurate. Four months before the 1882 convention, Ernest Morris had delivered 1800 orchids to Erastus Corning. It is entirely possible that, preoccupied with the care of his new plants, Corning dismissed the notion of running for Governor.

Before his death in 1897, Corning donated his beetle and butterfly collections to the New York State Museum. Today they still reside there, described as being in no worse condition than if they had been collected a week ago.[25] Many of the specimens were collected by Ernest T. Morris, but the entire collection was poorly documented and amateurishly cataloged prior to being received by the museum. Corning's two grown sons, Parker and Edwin, had the wherewithal to keep the Glenmont estate, which they divided between them. Both sons went on to build up successful businesses, and remained active in Democratic Party politics. Parker Corning sold the last remnant of the farm in the 1940s, and its renowned stock of Guernsey cattle were dispersed to dairy farms around Albany.

Erastus Corning Jr.'s prize orchids were auctioned to Mrs. George B. Wilson of Philadelphia. Mrs. Wilson became known as America's "Orchid Queen," and added to her collecting by sending her own orchid-hunters--George Barrault, George L. Freeman, and John F. Murphy--out into the rainforests of the world. As much as Ernest T. Morris sometimes played the adventurer, the orchid-hunters that came after him were rough-edged characters with little of the romantic sensibility and curiosity of the naturalist. It became known as a deadly, dangerous job; and those who survived their journeys came back with adventure stories fit for pulp magazines. By the 1920s, the age of the orchid-hunters had, for the most part, passed into history; orchid propagation from seed had taken the place of plants collected in the wild.

---

[1] "The Heart of the Andes," Brooklyn Daily Eagle, February 10, 1884, p.10.

[2] "Demerara," *Chicago Daily Tribune*, July 1, 1884, p.7.
[3] "A Captain's Story," *Harrisburg Independent,* April 18, 1884.
[4] "City News," *Indianapolis News*, April 21, 1884, p. 3.
[5]George Bowles, Morris's traveling companion to British Guiana, returned to Chicago and, by the late 1880s, was working as a theatrical agent. He did advance and press agent work for David Henderson, Eddie Foy, Alice Nielsen, and Frank Daniels.
[6] "A Distinguished Visitor," *Lebanon Patriot*, May 10, 1884.
[7] "College News," *Cincinnati Tribune*, May 12, 1884.
[8] "Local News," *Shelbyville Daily Evening Democrat,* August 20, 1884, p.4.
[9] "South America," *St. Joseph Daily Herald,* July 7, 1885, p.4.
[10] "Letter from Georgetown," *Galveston Daily News*, November 26, 1885, p. 8.
[11] Untitled news item, *Fort Scott Daily Monitor*, October 19, 1886, p.4.
[12] "Cobesanga and Copawassa," *Fort Scott Daily Monitor*, October 16, 1886.
[13] "In Search of Victoria Regia," *Fort Scott Weekly Monitor,* February 3, 1887, p.3.
[14] "Personal & Society," *Indianapolis Journal*, November 20, 1887, p.8.
[15] This plant was not named after Ernest Morris, but after Mr. Ernest Measures, a British orchid grower.
[16] "Death of a Collector," *Indianapolis Journal*, April 30, 1891, p.8.
[17] Ancestry.com. *Indiana Deaths, 1882-1920* [database on-line]. Provo, UT, USA: The Generations Network, Inc., 2004. Original data: Various Indiana county death records indexed by the Indiana Works Projects Administration. Indiana: circa 1938-1941.
[18] "Death of the Boy Traveler," *Chicago Daily Tribune*, May 9, 1891, p.9; "Ernest Morris' Life," *Washington Post*, May 3, 1891, p.12; "Death of Ernest Morris," *Brooklyn Daily Eagle*, April 30, 1891, p.4.
[19] The administrators of Crown Hill have no documentation that would explain why no markers exist on the Morris/Perry plot.
[20] "A Christian's Life Work," *Thornton* (IN) *Argus*, January 1, 1898.
[21] Morris, Sanford. Clippings scrapbook. Compiled 1875-1884. Privately held by Marianne Lanman Salaymeh.
[22] Tucker, Hugh C. *The Bible in Brazil: Colporter Experiences.* New York: Fleming H. Revell, 1902. p. 228.
[23] See: Rand, Edward Sprague. *Traditionary stories extant among the Tapoia Indians Inhabiting the Valley of the Middle Amazon in Northern Brazil.* Detroit: Geo. S. Davis, 1882; Rand's trips to New York are mentioned in: Clarkin, Franklin. "Stalking the Gentle Orchid," *Everybody's Magazine*, v.20, p.823. Rand's suicide is mentioned in: Hutchins, Meredith. "Henry L. Rand: Photographer and Visual Diarist," *Chebacco, the Magazine of Mount Desert Island Historical Society*, v. XI, 2010; p. 50-65.
[24] "Vanished, A Fortune," *New York Herald*, September 26, 1897, sec.5, p.9.
[25] The state of Corning's collection was attested to in an email exchange with NYSM curator Timothy McCabe.

12 RIVER OF ANONYMITY

In 1912, Theodore Roosevelt lost his bid to regain the presidency, leading the upstart Bull Moose (Progressive) Party in a failed effort against the Republicans and Democrats. It was a disappointing reversal for the fifty-four-year-old politician, but he refused to sit and sulk at his home in Oyster Bay, New York. A couple of months after the election, he received an invitation from a museum in Argentina to travel to South American and give a few lectures. Similar offers came to him from Chile and Brazil. Seeking both a physical challenge and a scientific purpose to his journey, Roosevelt consulted with his friends at the American Museum of Natural History. An itinerary was drawn up that would have him traverse the southern part of the continent before coming to the Amazon. The expedition would steam up the Amazon and Rio Negro to the Casiquiare, which provided a link to the Orinoco, and would then proceed to the coast of Venezuela. Roosevelt left the planning to others who had some experience with travel in South America.

Once onboard the ship heading southward, Roosevelt thought that the proposed route--although it passed through wild and remote areas, including down the cachoeiras of the Tapajós and up those on the Rio Negro--was well-known territory. Hearing this concern, a Brazilian foreign affairs minister rashly suggested that Roosevelt could attempt to map the course of an unknown river which only had been seen at its source by Colonel Cândido Rondon of Brazil's Telegraph Commission. The river had been named the Rio da Dúvida, or River of Doubt, since its path and outlet could only be guessed. Rondon was offered as co-leader of an expedition to the River of Doubt. To the consternation of many who believed the proposal was too dangerous, Roosevelt heartily agreed.

Roosevelt's American team had come prepared for a much different trip, stocking heavy provisions, lightweight canoes, and two heavy, custom-built motor-boats. Some of Roosevelt's team had experience traveling in South America, but none had attempted to bushwhack through rainforest, a skill that would be required to reach the River of Doubt. Colonel Rondon, fortunately, was the most experienced explorer Brazil had ever produced, and knew better than anyone else how to negotiate with hostile tribes for passage through their territory. Even with Rondon's skills, the expedition soon turned into a nightmare. Most of the nineteen men came down with malaria, one bearer murdered another and fled into the forest, and one drowned. Canoes were lost and crude ones had to be constructed. Roosevelt not only came down with a fever, he also suffered a gash in his leg that festered and caused blood poisoning. For days, he hovered near death. During much of their journey, the expedition was followed by the curious warriors of an uncontacted tribe, the Cinta Linga. The expedition ran out of food a third of the way down the River of Doubt. Fortunately, when close to starvation, they stumbled upon some rubber traders on the lower reaches of the river, who guided them to its outlet into the Rio Aripuanã, which itself joined the Madeira river.

Roosevelt survived the ordeal, but suspicions remain that it so weakened him that the experience hastened his death five years later in 1919 at age sixty. He was born two years after Ernest Morris, and there were many similarities between the two as young men. Both were fans of the adventure stories of Captain Mayne Reid. Both had a passion for natural history, and were avid field collectors. Both turned away from academic studies of botany, zoology, ornithology, etc., because they did not have the patience for that level of detail, compared to the physical immediacy of fieldwork. Both had a strong sense of morality; a lack of interest in commercial endeavors; and were both hailed as "progressive" examples of American manhood. Each used physical exertion to overcome the illnesses of their youth; and both likely hastened their deaths through the diseases they contracted in Brazil.

The comparisons end there. Roosevelt was born into wealth, and could have followed any career of his choosing. He chose public service, and became a public administrator, political leader, and volunteer cavalry commander. Ernest Morris had many talents, but ambitions for military and public service, leadership, and political acumen were not among them. Were it not for family responsibilities, he might have spent his last years continuing as a field collector, content to earn just enough to outfit himself for another journey. Roosevelt's single four-month expedition yielded a substantial accomplishment by which "explorers" are often measured: he helped fill in a blank spot on a map with the route of a geographic feature, a river. Soon after the expedition, the Rio da Dúvida was rechristened the Rio Roosevelt. For all of Ernest Morris's efforts, he could not make such a claim.

While Roosevelt was on the Rio da Dúvida and incommunicado with the civilized world, the *Indianapolis Star* tracked down Kate Perry Morris and asked her opinion of the Roosevelt Expedition. She did not have much to say, other than telling the reporter a summary of Ernest's travels. She knew that Ernest had not mentioned the River of Doubt; even the source had not been found by Rondon until long after Morris had died. In fact, the River of Doubt lay about a hundred miles southwest of the furthest southern point Morris reached on the Tapajós, so it was outside of any territory he traversed. With a hint of bitterness, Kate said she would await publication of Roosevelt's book on his journey once he returned; and would compare it against her husband's unpublished book.[1]

One by-product of the Roosevelt expedition was a report issued by the American Museum of Natural History entitled: "The birds of Matto Grosso, Brazil : a report on the birds secured by the Roosevelt-Rondon Expedition." Two of the three co-authors were the naturalists who had accompanied Roosevelt, but an additional co-author was one who had not been a member of the expedition: Herbert Huntington Smith.

After the 1879 publication of *Brazil, The Amazons and the Coast,* Smith had embarked on a five-year expedition to Brazil in May 1881--just a few weeks after Ernest Morris's returned from his fifth South American trip with the manuscript he had prepared for Scribner. Prior to leaving the United States, Smith indicated to newspapers that he would be journeying up the Tapajós, following it all the way into the province of Mato Grasso. In August 1881, it was announced that the provincial assembly of Pará had allocated $2000 to Smith to write a scientific report on the Tapajós.[2] However, for reasons never disclosed, Smith, his wife, and assistants never ventured up the Tapajós.

Instead, they spent five years amassing a huge collection of insect, bird, mollusk, mammal, and other animal specimens mainly in the provinces of Mato Grasso, Parana, and Rio Grande do Sul—but not the Amazon valley. Many of the specimens went to the Museo Nacional in Rio de Janeiro, but Smith also sent thousands back to the United States for sorting and classification. He was criticized by some Brazilian naturalists for exporting such a large quantity, although the right to do so was in his contract with the Brazilian government.

Smith spent the later years of the 1890s in Mexico and Trinidad, once again collecting specimens for Britain's Royal Society and other institutions. He returned to the United States to accept the position of curator at the Carnegie Museum in Pittsburgh, where he and his wife conducted many flora and fauna collecting field trips around Western Pennsylvania and West Virginia. His last South American expedition was to the coast of Colombia between 1898 and 1902. There he became seriously ill, forcing a return to the United States. After recovering his health, he led an expedition to study the fresh-water mollusks of Alabama. As a result of that work, he was invited to manage the museum of the University of Alabama. As Smith aged he became increasingly deaf, which led to an accident near the Tuscaloosa campus in which he was fatally struck by a freight train. He was sixty-eight.[3]

Ernest Morris's legacy represents a gray area in the longstanding debate of what constitutes "exploration" versus more self-serving "adventurism". Morris had no formal training as a scientist; no scientific instruments or mapping equipment; and no backing of any academic or geographic institution. He did not seek to chart any areas or rivers. He never published any scientific studies. Yet he spent most of the time on his journeys collecting specimens of flora and fauna, venturing into forest areas known only by his Indian guides. He was commissioned as a naturalist three times: by the Madeira railway contractors; by the Peruvian Steamship Company to explore the navigable reaches of the Napo; and by the provincial government in Iquitos, Peru to record the process used by the Yagua to make *wourari*.

By his accounting, Morris sent about a dozen plants that he did not recognize to botanists in Germany for analysis, but what the results were—if in fact the plants arrived in a condition allowing examination—is unknown.

His 1876-1877 journey up the Tapajós and Cururu rivers fit the label of adventurism. His goal was mercenary--to obtain Munduruku trophy heads to sell to collectors in the United States--and he knew that trade in these artifacts was illegal. He took great risks to reach the campos region; and he admitted that he smuggled the heads past Brazilian inspectors. However, he also took notes on the culture of the Campineiros while they still waged war against other tribes. Had his book been published, those sections on the Munduruku of the high plateau would have been cited for decades by the cultural anthropologists and ethnologists who later returned to that area. Morris's five remaining trips to South America were less foolhardy.

Other than the one trip, his ventures into the rainforest were not reckless. He traveled by steamer and canoe for the most part, and hiked deep into the rainforest only when there were native paths, often accompanied by a guide. He did not bushwhack miles through the forest, as Colonel Rondon and Percy Fawcett did many years later. Morris shunned entire river valleys--like the Madeira, the Tocantins, and the Trombetas--because they had reputations as disease hotspots. He aspired to ascend the Xingu and Japura rivers, but was dissuaded from those attempts due to reports of hostile Indians. More than once he hoped to venture across the Peruvian Andes to the Pacific coast, only to reconsider the idea of entering parts of Peru where war was being waged. He was not a thrill seeker, but did enjoy visiting remote Indian villages.

Morris could also be viewed as an agent of neocolonialism, paving the way for the exploitation of another country's resources. Ernest Morris's newspaper columns, journal entries, and manuscript are full of examples where he emphasized the presence in Brazil of resources that Americans could and should exploit: rubber, gold, diamonds, hardwoods, etc. He encouraged his own brother, Frank Morris, to set himself up as a rubber trader in Santarem, though Frank was not inspired to pursue this course.[4] The rubber industry period in Brazil represents a history of mistreatment, enslavement, and murder of indigenous peoples by traders, exporters, and the foreign markets they served. The legacy of misery from this period taints the motivations of all American and European travelers to the Amazon during this period. The explorers who charted paths of navigation for steamers or made notes on the economic value of natural resources were complicit in that abuse, even if they viewed it as "the inevitable march of progress." Morris mentioned that if Americans did not pursue those opportunities, the British and other European countries would do so in their place.

Morris himself exploited the Amazon's cultural heritage by trading for the trophy heads; and its biological heritage, by extracting thousands of plants, insects, and birds. He, as well as other naturalists and scientists who came before and after, saw themselves as serving the cause of expanding scientific and geographical knowledge. To their credit, Americans Herbert H. Smith and Charles F. Hartt conducted much of their field research and collecting of behalf of Brazilian institutions. However, cooperation with the Brazilian government of the 1870s and 1880s was a step removed from acts that can be construed as strengthening and empowering the native peoples of the country. It is difficult to argue that any of the foreign scientists that visited the "naturalist's paradise" in the nineteenth century transcended the cultural lens that assumed the superiority of their ethnic group over others. For that reason, today's cultural anthropologists tend to discount travel narratives from that period as flawed.

Today, many of the tribes of the Amazon basin that existed in Morris's day have died out or assimilated. More still are threatened with that fate. The Munduruku, with whom Morris was most familiar, are fighting against hydroelectric dam construction on the Tapajós that would flood much of their ancestral homelands; they also are fending off intrusions onto their land by loggers and gold prospectors. They are warriors once again, not for the sake of trophy heads, but as defenders of the forest.

Evidence of Ernest Morris's legacy is obscure, but exists. A Munduruku trophy head is still on display at the American Museum of Natural History in New York City. It was donated during the museum's formative period, and accession notes concerning it are incomplete, so documentation of provenance crediting Morris is lacking; but during his life he mentioned one of the heads was at the museum, and C. Hart Merriam also said two Morris heads were at the AMNH, so there is little doubt it was one of the ten he brought back to New York.[5] For nearly 140 years, museum visitors have been fascinated by this artifact, and have had their imaginations inspired to learn about the culture that produced it. The head that was donated to the Smithsonian by Dr. C. Hart Merriam in 1916 is also still in its collection, though not on public display.[6]

Morris's unpublished manuscript, *In Search of Human Heads,* was kept by family members, passed down by his daughter Enid to descendants of the Brown family and then the Lanman family. The book, along with journals, photographs, and the clipping scrapbook kept by Ernest's father, now are in the care of Ernest's great-great-granddaughter, Marianne Lanman Salaymeh. In 2016 the manuscript was transcribed and made available online via the Internet Archive non-profit library, and has now been viewed by anthropologists in Brazil and around the world.

Morris's *New York World* columns exist in digitized newspaper archives. Through those articles, which were often reprinted in major newspapers around the United States, Ernest Morris once reached more readers than any other nineteenth century American writer on the Amazon. The columns that Morris penned were better composed, more focused, and in many ways more evocative than *In Search of Human Heads*. However, they were never collated; and their ephemeral status meant that Morris was quickly forgotten once he stopped contributing columns. Joseph Pulitzer's new regime at the *World* did nothing to acknowledge the newspapers' past correspondents; they did not even mention Morris's death.

Ernest Morris's sole ambition was to be a naturalist, a vocation that he believed he was born to follow. The ideal he formed of the life of a naturalist harkened back to an earlier, late eighteenth century/early nineteenth century concept of "romantic science." As historian/biographer Richard Holmes notes in *The Age of Wonder,* "The idea of the exploratory voyage, often lonely and perilous, is in one form or another a central and defining metaphor of Romantic science."[7]

There is no question that Romanticism sustained Morris during his trips to the Amazon. It did so in a literal sense: he read books of Romantic poetry at nights in his huts, and composed his own verses. His friend Edward S. Rand was a noted poet. It was Morris's romantic sensibility that kept him going. His daily journals are full of moments of wonder: macaws of every color flying in all directions through the vivid green forest canopy; phosphorescent beetle larvae covering large termite mounds like stars in the night; butterflies flittering around a tree trunk like a kaleidoscope; canoeing around bends in a stream to encounter an orchid-filled scene—he lived for the beauty of those moments. He was neither a true scientist nor an adventurer; he was a Romantic observer of natural beauty.

Today, we recognize that a scientist who enjoys fieldwork can, with reluctance, withdraw back to the studious pursuits of collating, cataloging, documenting, theorizing, etc. Ernest Morris had no such alternatives. His travels were ended by a combination of factors: his responsibility to provide and care for his wife and child; his lack of capital; the withdrawal of support from the *New York World;* his frustration over not seeing his book published; the increasing difficulty of exporting plants out of their native countries; his frail health, i.e. bronchitis exacerbated by multiple bouts of malaria and other tropical diseases; and, perhaps, the lack of collector patrons. As Kate Perry Morris told the *Indianapolis Star,* Ernest could not endure living away from the Amazon: "He couldn't do it. That life meant everything to him.

---

[1] "Roosevelt Trail Known to Woman," *Indianapolis Sunday Star*, March 29, 1914, p. 16.
[2] "Provincial Notes," *Rio News,* August 24, 1881, p.3.
[3] Holland, William Jacob. "Obituary Notes." *Annals of the Carnegie Museum,* 1919, v.12, p.353.
[4]Frank Morris, Ernest's younger brother, after his sole international adventure in Brazil in 1878-1879, returned to Indianapolis and remained there, dying at age

89. He was married in 1880, a year prior to Ernest, when he was twenty-one years old. A son, born in 1881, was christened Ernest C. Morris. Frank's entire career was spent as a typesetter and printer.

[5] The most detailed account of the disposition of the eleven heads Morris obtained is found in "Old Reporter's Reminiscences," *Indianapolis News*, October 6, 1906., p.14. Journalist J. Cooley Fletcher relates that one of the heads was given to Captain Hoeffner, the Amazon steamer captain who helped Morris smuggle them to Pará, who sent it on to Berlin. Of the ten others: "Two became the property of the College of Physicians and Surgeons, Brooklyn, NY; two of the Smithsonian Institution; two of the museum in Central Park; and one or more sent to England."

[6] *Smoke Dried Head Full Size*. Catalog no. E289950; Accession no. 059457. Smithsonian National Museum of Natural History, Washington, D. C.

[7] Holmes, Richard. 2011. *The age of wonder: how the Romantic generation discovered the beauty and terror of science*. London: HarperPress.

Spelling Notes

Both "Munduruku" and "Mundurucu" are in current use as spellings of the Brazilian tribe. The spelling with the "k" is more prevalent, and for that reason alone has been used in this text.

In his writings, Ernest T. Morris used then-accepted spelling variations for place names that have since changed. To avoid confusion to the reader, I have altered the spelling of these names within quoted text to match current use. These include:

- Manaus in place of Manaos
- Juruti in place of Juruty
- Pebas in place of Pevas
- Tefé in place of Teffé

Other place names Morris used have since been changed to different names. These include:

- Lago Grande do Villa Franca is now known as Lago Grande do Curuai
- The city of Pará is now known as Belém
- San Antonio on the Madeira river is now inundated by the Santo Antônio hydroelectric dam. The city of Porto Velho is a few miles away to the northeast.

Glossary

Aldeia – Village. In Santarem of the 1870s, used to refer to the older, poorer section of town.

Anta -- *Tapirus terrestris*, the South American tapir.

Arara – Macaws; South American parrots

Bacaba -- *Oenocarpus bacaba,* a fruiting palm, used to make drinks; oil used for spider and scorpion bites; leaves used for thatching.

Barraca – A thatch hut

Bromelia -- type genus of the botanical family *Bromeliaceae*, subfamily *Bromelioideae*

Buzina – a horn made from bamboo stalks

Cachoeiras – river rapids; cascades, often signaling a drop in elevation

Capim -- river grass, found in large masses along the Amazon

Cacao -- *Theobroma cacao*; seeds from this small, wide-spreading evergreen tree are the source of cocoa, cocoa butter and chocolate.

Campineiros – Colloquial name for Munduruku peoples dwelling in the campos villages east of the Tapajós.

Campos – a tropic or sub-tropic grass plain of Brazil

Caramba – "Damn it". Portuguese.

Castanha -- *Bertholletia excels,* a tree whose seeds are known in North America as Brazil nuts.

Chicha -- A fermented drink made from maize or yucca, often after being mixed with saliva to break down the starches.

Cinchona – A genus of 24-40 species of trees limited to high elevations in the Andes from Bolivia to Columbia. Their bark contains quinine, used in the treatment of malaria.

Coaita – Black spider monkey. The archaic spelling Quatá persists in many place names.

Confederados – Veterans of the Confederate States of America army who chose to emigrate to Brazil following the end of the American Civil War.

Copaiba – An oleoresin used in laquers and varnishes derived from trees of the genus Copaifera. The oil is also used in folk medicines.

Cuia -- A container for liquids varying from cup to pitcher-sized, made from gourds of the calabash tree.

Curaca – Variant spelling of the Quechua *kuraka,* provincial governor; in use since the Inca empire.

Delegado – the head of a civil police district

Forno – a mud/brick fireplace used for cooking

Frei -- Friar

Furnas – grotto or whirlpool; also, used for cave

Gambá -- A single-headed, *barrel-*shaped *drum made by stretching a skin over the end of a hollow log.*

Garrafão -- carboy or large glass bottle protected by a wicker cover

Genipapo – A small tree, *Genipa Americana;* juice of the immature fruit is used to create indigo stains on human skin; used for tattooing by several South American peoples.

Igapó – Flooded forest

Igarapés – Streams; creeks

Igarité – A large working canoe, capable of supporting mast and sail, with a rudder and keel

Jararaca -- *Bothrops jararaca,* a Brazilian species of pit viper, responsible for more venomous bites than any other species in the Amazon basin.

Língua geral – an artificial language blending Tupi and Portuguese, devised to serve as a common language among peoples residing in the Amazon valley.

Maloca – A thatch longhouse used by Amer-Indian peoples of the Amazon basin.

Maquisapa – Black spider monkey; this name is used more often in the upper Amazon, as opposed to coaita on the lower Amazon.

Marfim-vegetal -- A common name for *Phytelephas aequatorialis,* a species of palm whose seeds are used as an ivory substitute.

Masato – A fermented beverage made from chewed yucca plants.

Mestiço – A person with both European and Amer-Indian ethnic heritage.

Montaria – A canoe made from a single piece of timber

Pacova – A common name for banana

Paraná-mirims – Water channels of sufficient size to allow navigation by a larger boat

Pium -- *Simulium pertinax,* a bloodsucking blackfly

Quartel – Headquarters; main building

Real – A unit of current in the Brazilian empire; in the late 1870s, Morris estimated its value to be $0.10 in American currency.

Remedio – Remedy (Spanish)

Remoinhos – whirlpools

Roga – A clearing made in the forest for agricultural purposes

Salsa – Colloquial name for sarsaparilla, woody vines of the genus *Smilax,* whose roots are used for medicinal purposes by peoples of the Amazon basin.

Tapiti -- A tubular wicker press used to remove moisture during the preparation of farinha and other starchy extracts

Teniente – Literally "Lieutenant" in Spanish; used in Peru for local authorities empowered under the authority of a provincial governor

Terra preta – Fertile, loamy dark soil found in spots in the Amazon basin; enriched by ancient Amer-Indians for agriculture

Tolda – Boat awning used to form a cabin, often made of thatch material

Tuxaua – Tupi language word for chieftan

Urucu – The Tupi language name for achiote (*Bixa Orellana),* a shrub with seeds that are crushed to produce a red pigment used in body paint.

Wourari – In Eastern Peru, a local name for the poison curare

Bibliography

Agassiz, Louis, and Elizabeth Cabot Cary Agassiz. 1868. *A journey in Brazil.* Boston: Ticknor and Fields.

Araujo, Ana Lucia. 2015. *Brazil through French eyes: a nineteenth-century artist in the tropics.* Albuquerque, N.M.: University of New Mexico Press.

Barber, Lynn. 1980. *The heyday of natural history, 1820-1870.* Garden City, N.Y.: Doubleday.

Bates, Henry Walter. 1962. *The naturalist on the River Amazons.* Berkeley: University of California Press.

Beard, Edward L. 1886. "The progress of orchid culture in America: read before the Massachusetts Horticultural Society, March 6, 1886". *Transactions of the Massachusetts Horticultural Society*, 1886. p. 123-151.

Britz, Billie Sherrill. 1977. *The greenhouse at Lyndhurst: construction and development of the Gould greenhouse, 1881.* Washington: Preservation Press, National Trust for Historic Preservation in the United States.

Chandless, William. 1866. *Ascent of the River Purus. Read before the Royal Geographical Society of London, on the 26th Feb. 1866.* London: William Clowes and Sons.

Craig, Neville B. 1907. *Recollections of an ill-fated expedition to the headwaters of the Madeira River in Brazil.* Philadelphia: J.B. Lippincott.

Edwards, William H. 1847. *A voyage up the river Amazon including a residence at Pará.* New-York: D. Appleton & Co.

Endersby, Jim. 2016. *Orchid: a cultural history.* Chicago: University of Chicago Press.

Goodman, Edward J. 1992. *The explorers of South America.* Norman: University of Oklahoma Press.

Goulding, Michael, Ronaldo Barthem, and Efrem Jorge Gondim Ferreira. 2003. *The Smithsonian atlas of the Amazon.* Washington, D.C.: Smithsonian Books.

Hansen, Eric. 2001. *Orchid fever: a horticultural tale of love, lust and lunacy.* London: Methuen.

Hartt, Charles Frederick, and Louis Agassiz. 1975. *Geology and physical geography of Brazil: scientific results of a journey in Brazil by Louis Agassiz and his traveling companions.* Huntington, N.Y.: R.E. Krieger Pub. Co.

Hartt, Charles Frederick, and Richard Rathbun. 1875. *Morgan expedition, 1870-71 ... on the Devonian trilobites and mollusks of Ereré, province of Pará, Brazil.* [Place of publication not identified]: [publisher not identified].

Hemming, John. 2015. *Naturalists in paradise: Wallace, Bates and Spruce in the Amazon.*

Hemming, John. 2008. *Tree of rivers: the story of the Amazon.* New York: Thames & Hudson.

Herndon, William Lewis, and Lardner Gibbon. 2000. *Exploration of the valley of the Amazon, 1851-1852.* New York: Grove Press.

Holmes, Richard. 2011. *The age of wonder: how the Romantic generation discovered the beauty and terror of science.* London: HarperPress.

Honigsbaum, Mark. 2003. *The fever trail: in search of the cure for malaria.* New York: Picador.

Hutchins, Meredith. "Henry L. Rand: Photographer and Visual Diarist," *Chebacco, the Magazine of Mount Desert Island Historical Society,* v. XI, 2010; p. 50-65.

Inlow, William. "The Indiana Physician as Geologist and Naturalist" *Indiana Magazine of History* [Online], (1 March 1960)

Irmscher, Christoph. 2013. *Louis Agassiz: creator of American science.* Boston: Houghton Mifflin Harcourt.

Jackson, Joe. 2008. *The thief at the end of the world: rubber, power, and the seeds of empire.* New York: Viking.

Keller, Franz. 1875. *The Amazon and Madeira rivers.* Philadelphia: J. B. Lippincott and Co.

Lange, Algot, J. Odell Hauser, and Frederick Samuel Dellenbaugh. 1912. *In the Amazon jungle; adventures in remote parts of the upper Amazon river, including a sojourn among cannibal Indians.* New York: G.P. Putnam's Sons.

Lange, Algot. 1914. *The lower Amazon; a narrative of explorations in the little known regions of the state of Pará, on the lower Amazon.* New York: G.P. Putnam's Sons.

Larson, Frances. 2015. *Severed: a history of heads lost and heads found.* New York: Liveright Publishing Corp.

Millard, Candice. 2005. *River of doubt: Theodore Roosevelt's darkest journey.* New York: Doubleday.

Muir, John, and Michael P. Branch. 2001. *John Muir's last journey: south to the Amazon and east to Africa: unpublished journals and selected correspondence.* Washington: Island Press/Shearwater Books.

Murphy, Robert F. 1960. *Headhunter's heritage.* Berkeley: University of California Press.

Murphy, Yolanda, and Robert F. Murphy. 1974. *Women of the forest.* New York: Columbia University Press.

Orton, James. 1870. *The Andes and the Amazon, or, Across the continent of South America.* New York: Harper & Bros.

Rand, Edward Sprague. 1876. *Orchids. A description of the species and varieties grown at Glen Ridge, near Boston, with lists and descriptions of other desirable kinds. Prefaced by chapters on the culture, propagation, collection, and hybridization of orchids; the construction and management of orchid houses; a glossary of botanical terms and significance of their names; the whole forming a complete manual of orchid culture.* New York: Rural Pub. Co.

Reid, Mayne. 1858. *The forest exiles, or, The perils of a Peruvian family amid the wilds of the Amazon.* Boston: Ticknor and Fields.

Rivière, Peter. 2006. *The Guiana travels of Robert Schomburgk, 1835-1844.* Aldershot, England: Ashgate for the Hakluyt Society.

Smith, Anthony. 1990. *Explorers of the Amazon.* London, England: Viking.

Smith, Herbert H. 1879. *Brazil, the Amazons and the coast.* New York: C. Scribner's Sons.

Spruce, Richard, and Alfred Russel Wallace. 1908. *Notes of a botanist on the Amazon & Andes: being records of travel on the Amazon and its tributaries, the Trombetas, Rio Negro, Uaupés, Casiquiari, Pacimoni, Huallaga and Pastasa : as also to the cataracts of the Orinoco, along the eastern side of the Andes of Peru and Ecuador, and the shores of the Pacific, during the years 1849-1864.* London: Macmillan.

Sterling, Keir B. 1977. *Last of the naturalists: the career of C. Hart Merriam.* New York: Arno Press.

Tully, John A. 2011. *The devil's milk: a social history of rubber.* New York: Monthly Review Press.

Wallace, Alfred Russel. 2010. *A narrative of travels on the Amazon and Rio Negro: with an account of the native tribes, and observations on the climate, geology, and natural history of the Amazon.* Cambridge [England]: Cambridge Univ. Press.

Wickham, Henry Alexander. 1872. *Rough notes of a journey through the wilderness from Trinidad to Para, Brazil, by way of the great cataracts of the Orinoco, Atabapo, and Rio Negro.* London: W.H.J.

Worster, Donald. 2008. *A passion for nature: the life of John Muir.* Oxford: Oxford University Press.

Wulf, Andrea. 2016. *The invention of nature: Alexander von Humboldt's new world.* New York: Alfred A. Knopf.

Made in United States
Orlando, FL
05 November 2026

90224854R00128